#fibre(S)cement.com

120 years of the history of an industry

Jacques ROULLAND

#fibre(S)cement.com

120 years of the history of an industry

Proofreading by Angela WILSON

Éditeur : BoD-Books on Demand
12-14 rond-point des Champs-Élysées, 75008 Paris
Impression : Books on Demand, Norderstedt, Allemagne

ISBN : 978-2-322230068
Dépôt légal : avril 2021

CONTENT

	Prologue	9
I	Invention and first developments	15
II	Asbestos	35
III	Manufacturing process and products	51
IV	Worldwide outburst	97
V	The crisis	127
VI	Fibre(s)cement, Art and Architecture	149
VII	Wind of change	181
VIII	Countries overview	195

Algeria 196, Australia 196, Austria 199 , **B**elgium 202, Bangladesh 203, Bolivia 203, Brasil 204 , **C**anada 218, Central-America 222 , Chile 224, China 224, Colombia 227 , Cuba 227, **D**enmark 233, **E**gypt 234, Ecuador 235, **F**rance 236, **G**ermany 238, **I**ndia 240 , Indonesia 242 , Iran 243 , **I**taly 246, Ivory-Coast 250, **J**apan 251, , **L**uxemburg 255, **M**alaysia 255 , Mexico 256 , Morocco 269 ,**N**igeria 270 , The Netherlands 272 , **P**eru 274 , Philippines 275, **R**ussia 276, **S**outh-Africa 280 , Spain 281 , Sri Lanka 287, Switzerland 289 , Syria/Lebanon 292, **T**aiwan 296, Thailand 296 , Turkey 299, **U**nited-Kingdom 301, USA 304, **V**enezuela 316, Vietnam 319, **Zi**mbabwe 322.

	Epilogue	325
	Thanks	329

PROLOGUE

'The ultimate aim of enterprises is -to be and to preserve being-. Doing such, they share hazards and perils. People in charge remain in the obligation of keeping with their continuity, which is finally their unique purpose'.

Roger Martin 'Patrons de droit divin' 1954

#fibre(S)cement.com Why to write this book?

It's 1963, I just joined the company as an assistant to the sale's administrative manager. This medium size enterprise was located in Angouleme, Charente[1], France, manufacturing felts[2] for the paper industry, the fibre-cement industry and a few other industries. The company belonged to a big French group named CGE. Exports were not much over twenty percent.

One of the first task I had to deal with, consisted in managing the 'temporary importation'[3] of two felts for asbestos-cement, which an Israeli customer claimed their

[1] Charente: region of the south-centre of France, mostly known for its Cognac.

[2] Felt: in the present case, a felt is a kind of large and wide belt, made of wool, cotton and/or synthetic fibres, used mostly in the paper industry. Its main purpose consists in participating to extract water out of the liquid paste in order to become paper. Its main characteristic is to be known exclusively by its manufacturers and its users...

[3] A temporary import in this present case, is used to import the good in its original country for a limited time, in order to correct or repair it.

length is not conform to ordered size. I wondered what could be the role of a felt in the manufacturing of corrugated asbestos-cement sheets...
Soon after, the French Société de Pont à Mousson, built a plant for asbestos- cement sheets and pipes in Valladolid, Spain. Our company received orders of felts for the start of the machines. Those bound to the sheet machine appeared to be somewhat short in length. A young Spanish engineer hurried up to Angouleme with three of them piled up in his Renault R8. We had to lengthen them immediately...
At the same time, I heard the story of our Export Sales Manager who, a few years earlier had been travelling to Brazil for a prospecting journey. On the banks of the Amazon in Manaus, or perhaps in Belem, he had met a fellow countryman in charge of an asbestos-cement manufacture, belonging to the same French group which also owned the factory in Valladolid we have just been talking about.
Indeed, asbestos-cement questioned me more and more every day. Pont-à-Mousson was also partner of an asbestos-cement company near Düsseldorf, Germany. Our Export Sales Manager had to travel there. As he did not speak German, he did not feel at ease and he asked me to accompany him. I was not enthusiastic as, in spite of all my previous efforts, I was not too good at Goethe's language, and I am not better now...
Two years later, our Export Sales Manager retired. Our President suggested that I replace him near our 'asbestos-cement' customers in France, Spain and Germany. Such a proposition sounded like a pretty good opportunity to improve my position. I did not hesitate in spite of a certain concern from my wife who already imagined my future absences.
From this day, I was involved in a kind of adventure which lasted more than thirty years, and brought me to develop our sales, first among the French and foreign branches of the famous Group Pont-à-Mousson[4]. Later on, with lots of other companies of all sizes through Europe, Africa, Latin-America and South East Asia.
The consequence is that I visited lots of countries, lots of manufactures and above all, I met plenty of people, who loved their job and did their best to make their companies prosper. We would often meet again from one factory to another, with different climates and different working conditions. We often converted into friends.

---

[4] People working with **P**ont **a** **M**ousson, used to call their company from its initials: PAM. If you don't mind, I will use this abbreviation every time I will have to refer to this venerable company.

Prologue

Our children were growing-up and we talked about them while sharing meals sometimes in good restaurants, sometimes in cheap ones.
Time to retire arrived. The *thirty glorious* were already far behind us. Asbestos had caused deaths among workers and also among factories. Many of them were closed, leaving brownfields making our suburbs ugly. With free time offered by retirement, heavy eyes glanced toward old days. The desire to write the history of what I had seen and lived became a kind of moral obligation.
Then, after some fifteen years away from felts and fibre-cement factories, I started looking for my former contacts in France and abroad. I found out that everything had changed after the big crisis resulting from the *asbestos scandal*, but that a new range of products had been developed. Even, if our country remains very dubious about these innovations, I understood that in spite of all the faults affecting fibre-cement, it offered roofs and often water supply, to millions of people worldwide. This alone certainly justifies writing its memory and to show that, even after a drama such as the *asbestos scandal*, a flourishing but destroyed industry could pick itself up again. Like a country reviving after a war and rising out of its ashes.
An evidence imposed itself in my mind. The life of an industrial product is quite similar to the life of a human being:
-Invention: Conception, pregnancy.

-First industrial manufacturing: Birth.

-Development of manufactures and sales: Teenage years.

-Globalization of productions and sales: Adult age.

-Crisis due to death hazards: Illness.

-Creation of new technologies: Remission- Recovery.

-New developments after recovery: Adult age continuing, without knowing for how long, similar to old age for the writer and his readers.

As a conclusion to this prologue, let me point out that one will not find hereunder any detailed description of processes developed by fibre-cement manufacturers. Experts did it before me and much better than I could do. Along the following pages, I will do my best to develop the historical and human aspect of the subject, while mixing anecdotes and casting a glance to existing links between art, architecture and fibre-cement.

The reader should not expect, neither a plea against asbestos, nor against its users. In this matter many skilful people worked and published[5].

My contacts with industrials in charge of fibre(s)-cement, confirmed that they all worked respecting norms applicable in their respective countries. Though strongly criticized, this industry, by far, is not the only one to make our lives hazardous. On the very days when I write, not only some countries still refuse to admit the dangers of asbestos, but a very great part of the global agriculture and farming does not hesitate to poison us slowly.

An overview of troubles and issues met when intending to keep alive such an industry for decades appear through the few following lines. They belong to the speech given in 2006, by *Alain De Metz*, high executive of Saint-Gobain Group, on the occasion of the official opening of a new factory at the border between Mexico and USA:

'Talk about this so disputed subject by now, is not an easy task, neither is it to talk about the development of the new offers of building materials. My short message refers to, or announces, a time which seems very near to those among us who lived it and still live the different stages of this fibre-cement activity. It was born in Mexico in the 1930/1940, and earlier in Europe and other parts of the world. For quite a lot of us, it will already be "history", for some others just a simple reference.

During the fifties and sixties, named "glorious", fibre-cement raised incredible tops. There, young people of the time found emotive sources, mixing the know-how of elder experienced people with the efforts of young engineers and workers. Each of

---

[5] Readers interested in illnesses linked to asbestos can revert to www.helsinki.fi/iehc2006/papers2/Hardy.pdf or to *"Eternit et l'Amiante 1922-2000"* by Odette Hardy-Hémery, published at Presses Universitaires du Septentrion. Another source of information can be found in: "AMIANTE 100 000 MORTS A VENIR, by François Malye, Le cherche midi ,2004.

them was participating to the birth of new technics which permitted to slide gently from almost craftsmanship productions, to mechanical processes. At the same time, they were keeping alive the knowledge and sensibility of foremen already growing older. This know-how from our elders certainly was the most prized of our wealth...'
The inquiry work required to write the present book turned out to be a second adventure, which I did not imagine when I began. It led me not only to rediscover many of my former contacts, but also to discover new people from Brazil to China, sometimes via Canada...
Unfortunately, due to the asbestos tragedy, some of my potential interlocutors did not wish to talk. Some even refused to meet me, and a few of them are now dead.

If you are not familiar with the world of fibre-cement, some purely technical and historical chapters might be quite boring. I suggest reading: Prologue, Chapters I and II, then jump to Chapter VI, you will discover quite a lot of unknown topics, skip to Chapter VII, browse chapter VIII, and choose the countries that you find interesting.

I Invention and first developments

"A new invention which will certainly revolutionize every roofing system, was recently patented in Austria"
Cirkel[6]

At the end of the 19th century, in industrialized countries, intensive research was progressing in order to invent light and profitable products intended for the building industry.

Around 1892, *Kuhlewein*, a German, was blending asbestos and cement with the aim of achieving light and non-inflammable roofing[7].

In the preface of the book dedicated to *'Fibrecement in Arts 1903-1973'*, published by Société du Fibrociment-Elo in 1973, *Guillaume Gillet*[8],informed:

"*Asbestos-cement dates back to the end of the last century. Around 1892, the German Kuhlewein blended asbestos and cement in order to obtain non-inflammable roofing. In 1896, a German-Russian manufacturer named URALIT, was producing some sheets which generated a few applications. In 1898, SIMMONS & BOCK, were making asbestos-cement tiles with small quantities of water*".

---

[6] Cirkel, Canadian writer of *Asbestos, its occurrence, exploitation and uses*. Department of Interior Canada, Ottawa, 1905 page 136.

[7] *R.A. Weniger*: *die Asbestzementschiefer Fabrikation*, Berlin 1926.

[8] *Guillaume Gillet*: famous French architect of the 20th century, designer among others of Notre Dame in Royan, of Palais des Congrés Porte Maillot in Paris. Premier Grand Prix de Rome, Chief Architect of French Civil and National Buildings.

One will note that the name *URALIT* can be found in several places, sometimes with some variation, all along the history that we begin to tell. One can consider that *Ural*, where asbestos was mostly produced at the time, and the Greek suffix *ite,* standing for *natural stone,* clearly justified the spread of the name.

In the US in 1897, Mr *Norton*, an assistant at the *Temperatures Measures Department of M.I.T.*[9], was invited to work on researching some kind of replacement material to wood, especially in order to reduce fires. He worked on the project until 1903. He was supported by powerful companies of the country, and he managed to conceive a mix of cement and asbestos looking like mortar. He would drop the mortar on a filter, then press it very hard, which enabled to obtain homogenous sheets of different thickness. In 1905, *M.I.T.* filed patents and *T.F. Manville*, a company that we will meet several times, among others, began to produce sheets. Yet, the new process was not adequate to obtain thin sheets, necessary to make artificial slates.

Americans carried on searching and happened to finalize a second Norton process constituting in the deposit of a thin homogenous layer of asbestos and cement mortar on an endless cloth. The resulting sheet was dried and highly pressed between two cylinders, then slates could easily be cut. The process was patented in 1910 and successfully used for many years.

In the beginning of the 20th century, US industrials finalized a mixture of *Portland* cement and asbestos fibres and obtained a lasting and fireproof building material.

Manufactured for the first time in 1905, this new product was used to coat boilers, steam pipes and any type of device producing intense heat. Ceilings and roofs including hot pipes were very often equipped with such a compound.

However, the actual father of Asbestos-Cement, worldwide recognised, is ***Ludwig Hatschek*** who invented the manufacturing process which was going to become the. International standard.

During the last years of the 19th century, *Fritz Hatschek* owned a cardboard factory producing asbestos reinforced cardboard in Vöcklabruck, Austria. Part of the cardboard was used as coasters.

---

[9] M.I.T.: Massachusetts Institute of Technology, famous private research university in Cambridge, Massachusetts, USA, founded in 1861.

One of the workers noted that cement wastes and drops of water would give birth to some kind of *'reinforced blend'* when mixed with asbestos fibres coming from the coasters of the *café*. *Fritz Hatschek* told the case to his son *Ludwig*, recently engineered. The latter started research, mixing different types of fibres. Soon he happened to obtain a kind of *'thin and stable reinforced cement'* which did not vanish against wetness.

Indeed, he was using the method already developed in cardboard machines, operating with rotary cylinder sieves held in *vats*, also named *sieves- cylinders*. He had the idea of mixing a highly diluted suspension of cement with carefully defibred asbestos in those tanks, in a very similar method to the one already used in pulp paste to make cardboard.

01-001. Ludwig Hatschek. (Uralita: 80 años).

Thanks to the rotary system and a wire *(a felt)* just like in the paper and cardboard industries, a thin skin *(a layer)* of paste was laid on a cylinder, which was also rotating. When after several rounds of the cylinder, the thickness was sufficient, he would cut it and let it fall on an exit flat conveyor. A kind of sheet of wet consistency was then obtained, very similar to fresh cardboard. *Ludwig* found out that the best results were reached with *Portland* cement. The wet sheet could easily be shaped, but once the cement hardening was set, the same sheet turned into a very solid board.

Then, the process only needed to be adapted to fit the requirements of the building industry: flat or corrugated sheets, artificial slates or cladding *(US: siding)*. *Ludwig* held a technique that was going to grow worldwide and which is still in use in the 21st century, after lots of adjustments in paste components as well as in machinery.

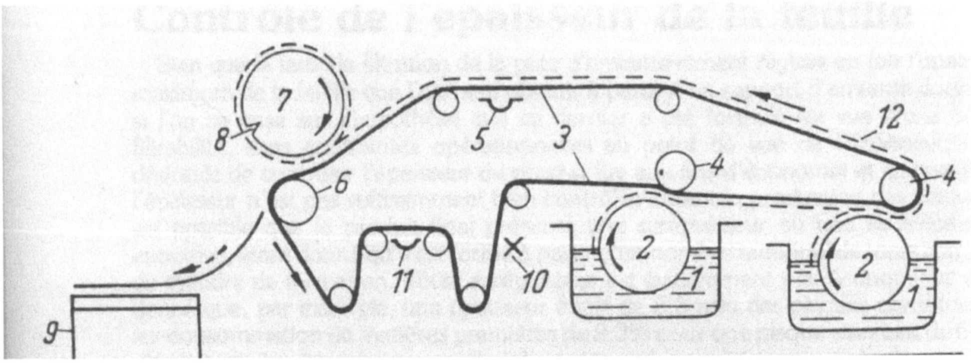

1.Slurry vat. 2.Sieve cylinders 3.Running felt. 4. Couch rolls. 5.Vacuum box.
6. Reward roll. 7.Formation Cylinder. 8.Cutting wire 9.Take-off conveyor.
10.Whopper.11.Suction box. 12.Sheet

01-002. Scheme due to InspectaPedia and JP Guerber

On March 27th 1900, he filed for a patent. Authorities were reluctant as the process of diluting cement with 100% of water was in complete opposition to the approach of the time. He was first denied in 1902 and finally he obtained the desired patent in 1905[10]. This delay did not stop the start of industrial production. Soon the new product received his patented name: *ETERNIT*, supposed to allude both to durability and to stone. Then, the old *Köchmühle* cardboard mill, near *Vöcklabruck* Austria, gave birth to: *ETERNIT WERKE LUDWIG HATSCHEK,* whose first part of the name was going to be famous worldwide.
In the beginning of the 20th century, European-Union, high speed trains and low-cost flights did not exist, but travels inside Europe were very frequent, especially among bourgeoisie and ambitious industrials willing to develop their business. Luxurious international trains dotted with comfortable sleeping-carriages and rich dinning-cars

[10] Details due to an article of H. FREY in « Bulletin Technique de la Suisse Romande »

helped easy journeys between big European towns. Passports where seldom required. Gold money avoided exchange and converting trouble. Managers did not yet need to justify their expenses with *HRMC*[11] *(US: IRS)* nor with their share-holders or employees.
That is how innovations were soon known from one country to the other, all the more as undertakers were free to build new premises, to hire workers, and to offer new products on the market, without fighting to obtain licences from any agency or from numerous and all-powerful organizations.
In such conditions, *Ludwig Hatcheck's* invention soon escaped from Austria, expanded all through Europe and even to some other continents. *Ludwig* owned his patent, when industrialists put forward their interest for the new product, he would grant the licence, limited to one per country, on condition that the name *ETERNIT* be maintained by the new producer[12].

01-003. First Hatschek Machine. Web Voith, St. Pölten, Austria

[11] HRMC : Tax collection agency in UK
[12] Massimo GIGLIATTI : « ETERNIT, Histoire d'une Entreprise Encore Active ».

On the above picture, the fact that the felt be not put on the machine, helps see vats, rolls and the important pipe system needed to make the machine work.
Fast the henceforth famous name thrived almost everywhere to such a point that in many countries, it became synonymous of asbestos-cement. Indeed, the rule of *one only name*, demanded by the inventor soon knew exceptions as we will discover afterwards.
Visitors were rushing to *Vöcklabruck*! The first one was a Swiss, named *Alois Steinmann*, who acquired the patent and created *the Schweizerische Eternit AG* in *Niederurnen, Glarus, Switzerland*. Over the years, it became one of the biggest companies in the world in its speciality, first through exports and soon by creating multiple branches worldwide. The area of the first factory was elected on account of the high number of unemployed persons due to the, already existing, textile crisis.
Let us have a short break and check the ability of undertakers starting-up this so promising industry.
In 1905, a Belgium, *Alphonse EMSENS*, created ETERNIT in Belgium and opened a first manufacture in *Haren*, near Brussels. Linked to others, especially to his Swiss homonymous, this new company also gave birth to a worldwide giant. As soon as 1910, *Ernst Schmidheiny* from Switzerland organised a cement cartel. In 1930, thanks to *Holderbank Financial Holding,* he succeeded in including most of the cement makers of the world under his control. At the same time in Belgium, *Emsens* already mentioned, achieved an almost exclusive licence for heighten years on the British market. There, he gave birth to *G.R. Speaker* who imported asbestos-cement products from Belgium. He succeeded in maintaining such an activity until around 1975, excepted during the two World Wars. In 1922, *Ernst Schmidheiny* and *Alphonse Emsens* met for the first time. Then, both powerful industrialist families began a cooperation which was going to last some seventy years, covering cement, asbestos, and asbestos-cement markets.
During the same year 1922, the French *Joseph Cuvelier* and the Belgian *Jean Emsens* came closer together. They agreed to open an asbestos-cement factory in *Prouvy, Nord, France*. Production began immediately and the first *diamond-shaped-slates* designed by *Hatschek*, inspired Eternit logo now worldwide known.

01-004. Famous Eternit logo. Web Eternit. diamond-shaped-slates designed by *Hatschek*, inspired Eternit logo, now worldwide known, and still present in our 21st century.

In 1929, the Swiss *Ernst Schmidheiny* and the British multinational *Turner & Newall (T & N)* created the *Cartel International Asbestzement AG*[13]" known as "SAIAC". In 1985, the Swiss author *W. Carina* would write: '*After the war, Schmidheiny began to reproduce with asbestos-cement manufacturers what he had already performed with cement and brick makers. That was: gathering them together in a cartel*[14]".

The different "ETERNIT" companies from Germany, Switzerland, Spain, France, Belgium and Italy took on a share of "SAIAC" and set the head office in *Niederurnen, Switzerland*, where the *Schmidheiny* family was based. The Belgian received 23%, the Swiss received 27% and the balance was divided between the others. Close relationships also existed between SAIAC and the biggest manufacturer of asbestos-cement in USA, *John Manville (J-M),* who also owned the biggest asbestos mine in Canada. *John Manville* was also owner of 10% of Eternit Germany. The *English T & N* owned large interests in the asbestos mines of South Africa. In his yearly report 1929, *T & N* explained his decision to get involved in the cartel with the following words:

'We have become such a large part of the Nation's entire industry, that we have been able to arrange an International Cartel with the main manufacturers of ten European countries. The position of the European Asbestos Cement Industry is thus rationalised, and we expect great benefit by way of improved technique and economy to accrue to all concerned. This miniature League of Nations has a great future before it, for it is based upon the principle of mutual help, which now displaces the previous atmosphere of distrust and suspicion'.

The British government looked on such agreement with a favourable eye, considering that it participated to the nation's lifeblood. One can legitimately imagine that

---

[13] *International Asbestos-Cement Cartel.*

[14] Reported by *Bob Ruers in* ETERNIT AND THE SAIAC CARTEL. *Bob Ruers* is a Dutch lawman and senator, specialized in asbestos problems, who published *Asbestos Dynasties-the Eternit Multinational.*

many members and managers of other involved countries were sharing the same view.

One must note two very important points of the pact: -

-Setting up in Switzerland a research institution common to the entire industry.

-An agreement to set up factories in countries regarded as being neutral.

These two points highly favoured the ascendency of Switzerland on the market worldwide during the following decades. Let us again quote the Swiss W. Carina: *'The international Eternit scene of the 1920s and 1930s resembled a clan in which some members were married to each other, while others were related or had become friends as a result of their common interests'.*

This did not prevent low-blows between members when necessary[15]. During the fifteen following years Eternit Switzerland extended its position inside the cartel and won more or less one third of the total. At the same time, they also won a very strong position in the mining activity, vital for the supply of their asbestos-cement factories. The *Schmidheiny* family, together with the American *Armand Hammer*, even obtained from the Soviet Union, the concession of a mine in Ural. The whole of those events granted Eternit a very strong position in the asbestos-cement business. They had no exclusive rights, but a widely first position among rich countries, as well as in those with a large promising future[8].

During the first years following the invention, also appeared a French, *Edmond Oscar Lanhoffer*[16] from Mulhouse, near the border with Germany and Switzerland. He did not care about the rule settled by *Ludwig Hatschek* regarding the name Eternit and as soon as 1902, that is before the patent got granted, he set up *Fibrociment de Poissy,* in Poissy, some 45 km west of Paris. The new company soon knew a fast and important development. We will revert to this company when talking about the strong links existing between fibre-cement, art and architecture.

Immediately also appeared an Australian, who recently immigrated from UK. He was looking for industrial opportunities to develop in his new country. During a journey to Europe, he discovered asbestos-cement, it seems in France, and he decided to import it to Australia. Soon his sales met a great success, which, years later, gave birth to a new giant of this industry. He was *James Hardie*, whom we will talk in details later-on.

---

[15] Some historical details are also due to Série J. from Archives de Poissy, 78.300 France.

[16] Olivier DELAS « Les Établissements Industriels du Fibrociment de Poissy ».

As soon as 1906, in *Casale-Monferrato*, near Milan in Italy, handmade pipes were produced with manual welding. They were used as guides for rain waters flow down from gutters or as ventilation ducts. On his side, Ludwig Hatschek was searching to manufacture pipes without welding. His death in 1914, did not give him a chance to succeed.

In 1912, in *Widnes, Halton County, Cheshire, UK*, two companies named Everite[17] and Asbestilite, were the first ones to manufacture asbestos-cement in Great Britain. Reading *the Official Handbook of Widnes, from 1920*, one can note:

"It is a curious short-sighted policy, while we spend millions annually in grappling with fire, we continue almost invariably to construct our buildings of highly inflammable materials, when in Everite we have a building material which the fiercest fire cannot pass".

In fact, the honour of conceiving an industrial process for the manufacturing of pipes, belongs to *Adolfo Mazza*, founder of *Eternit Italia.* In 1912, he succeeded in avoiding the welding. His invention permitted a huge development of all kinds of asbestos-cement pipes, including pressure pipes which invaded the world.

01-005. Adolfo Mazza, father of the pipe machine of his name.(Uralita 80 años).

Before being cut and pulled apart by the forming cylinder, an asbestos-cement sheet looks like a pipe, with a rather big diameter and a short length. Pull it apart without losing its rotundity sounded somewhat tricky. Alfonso *Mazza's* genius idea

[17] The name « Everite », is independent of Everite France and Everite South Africa, which we will find out in different occasions in the present book. These two are also independent one from the other.

was to add a compressing device equipped with hydraulic jacks and to find out how to reduce the pressure progressively. In 1922 and 1923, he filed two patents about his invention.. Of course, he had to adapt length and diameter of the *make roll* to the desired sizes of the pipes to manufacture. This obligation committed him to in-

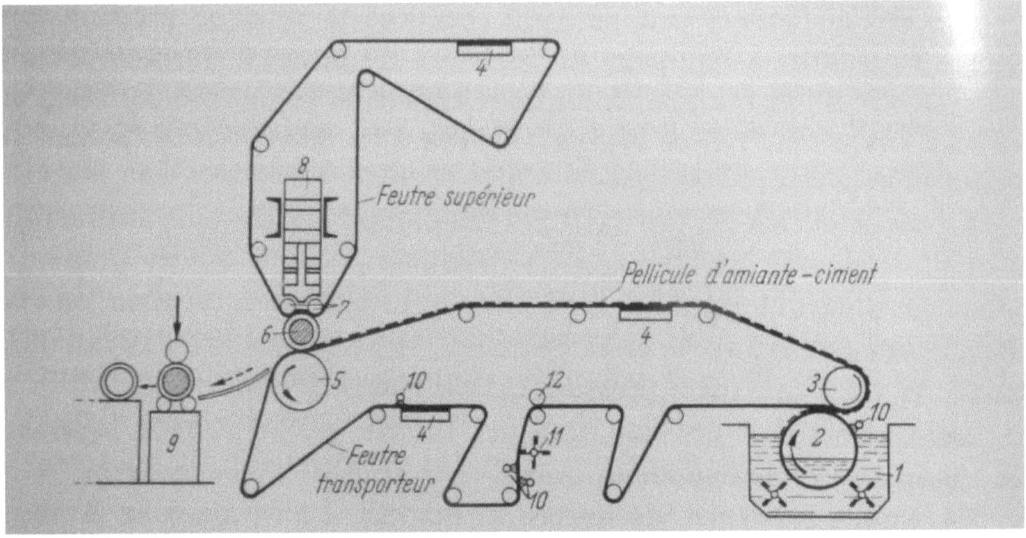

1: vat. 2: sieve cylinder. 3: couch roll. 4: suction box. 5: motor roll. 6: mandrel. 7: pressure rolls. 8: pressure device. 9: calender. 10: water spray. 11: felt beater. 12 : dewatering- rolls.

01-006 Mazza Pipe Machine. Kurt Hünerberg

crease the width of the machine and to reduce the diameter of the *making roll*. Later this became a *mandrel*, that is a kind of removable moulding cylinder or inner mould, which could be extracted once the pipe had reached a sufficient thickness and correlative strength. Interesting details and explanations can be found with InspectAPedia. During the first years, length was limited to 3 metres (39in) and diameter to some 20 centimetres (7,87in). Over years, length and diameter increased due to requirements of the market. This generated a gigantic development of asbestos-cement pipes applied to multiple uses. We will have opportunities to revert to the matter.

In the southwest of France, our Spanish neighbour was also very active. In Cataluña, an eminent businessman, *Don José Roviralta*, became aware of the qualities of the revolutionary new building material. He got in touch with *Ludwig* and *Adolfo*. As soon as 1907 he created a first workshop in *Cerdanyola del Valle, Cataluña, Barcelona, Spain.* Three years later, the company took the name of *URALITA* (it seems that the name was also built from Ural -and from the Greek 'ite' already met, turned into "ita" according the method dear to the heart of our Iberic neighbours, unless it be a reminder of the last letters of the name Roviralta)[18]. Along the 20th century *Uralita* also became a giant in the building material in Spain. We will also revert to this company later on.

In 1910, Russia turned to be interested in the new product. Czar *Nicolas II* ordered three fibre-cement machines from the Danish *F.L. SMITH.* They were not yet delivered when the 1917 Revolution arose...but one of them was not lost for everybody...We will talk about.

In 1912, the *Schweizeriche Eternit Werke AG* opened a new factory in the Netherlands. At the same time, production was turning at full capacity and exports knew an impressive rise towards Africa, Asia and Latin America.

In 1913, in *Grignoni* near Trieste, Italy, an environmental local agency appealed *Eternit* to courts for "damage to landscape"[19].

In 1914, Eternit Switzerland undertook to produce prefabricated buildings which generated the hire of many architects. In France *Fibrociment de Poissy* began to produce inner decorative panels.[20] The beginning of WWI played a part sometimes positive in the evolution of the asbestos-cement industry. For instance, Eternit Switzerland created branches in France, Great Britain and even in Australia. On the front, asbestos-cement helped to build shelters, back lines command posts, hospitals, warehouses, aircraft sheds, and even huge sheds for aerostats up to 35 meter long.

---

[18] In the suburbs of Toulouse, South west France, also existed at the time a company named URALITE, which was equally producing asbestos-cement, and of which *Don José Roviralta* was also the main shareholder. In Higham, Kent, UK, also existed a company named URALIT. The author did not happen to find out whether it had any link with the other ones wearing similar names.

[19] *Massimo Gigliatti : Eternit, Histoire d'une Entreprise encore active.* (History of a still active Company).

[20] Olivier Delas: "Les Etablissements Industriels du Fibrociment", *Chronos Poissy Octobre 2013, page 5.*

Some people claimed that users did not need coal as asbestos took care of heating them... In that state of war, in 1917 negotiations were going on, we don't know be

01-007. Asbestos-cement aerostat shed in Ecausseville Manche France.

tween who and who[21], about the installation of an asbestos-cement factory in *Outreau, Pas-de-Calais, France*. The location was considered as being too near to the front line, and the new factory was finally settled in *Bassens, near Bordeaux, Gironde, France*

[21] We can only suppose that these negotiations were held between the French War Ministry and Eternit...)

01-008 Advertising pavilion. Collection O Delas

01-009. Header section of Société Française de l'Everite, in 1925. Private archives

Two inhabitants of the small town of Lormont, next to Bassens, one of them being a bricklayer, then created *Société Française de l'Amiante-Ciment* (Asbestos-cement French Company) which soon became *Société Française de l'Everite* and just *Everite* in 1938. Let us note the always present concept of duration: *ever* and stone: *ite*.

At the same time, *Joseph Cuvelier*, sugar industrialist established in Northern France, was interested in building material market driven essential due to this rebuilding period. First, he started with asphalt cardboard. Soon after, he heard about the new product and immediately undertook a study trip to Austria, Czechoslovakia, Germany, the Netherlands and Belgium. When he came back home, he was convinced about the future of asbestos-cement and he purchased a 67 000 square yards plot, ideally located in Prouvy, Nord, France, and he created *Société des Matériaux Spéciaux de Construction* (Special Construction Materials Company). In 1922 he heard that the *EMSENS* family in Belgium was manufacturing asbestos-cement in the town of Haren and that they wished to set-up a branch in France. He contacted them and they both agreed to operate the Hatschek Austrian patent under the name of Eternit. As soon as October of the same year, the first sheet machine was working in the manufacture of *Prouvy.* A second one was in operation from January 1923[22]. *Joseph Cuvelier* withdrew from asphalt cardboard and devoted himself to asbestos-cement.

Production grew very fast due to the dynamism of the young company. Indeed, the post war situation of this totally devastated area was very favourable for such activity. Eternit made the most of the 'war damage compensation' resulting from the destruction of the old sugar refinery which they did not rebuilt. Such type of financing was quite frequent at the time[23].

In Belgium, two partners, *Leon Scheider* and *Camilla Van Kerkhove*, specialized in building industry for roof and frontage, also got into the manufacturing of asbestos-cement. They created S V.K.[24] in Sant-Niklaas, company which is still operating when I am writing these lines.

Let us revert to France for a while:

In 1924, Société de Pont à Mousson gave birth to Société Parisienne des Procédés Hum, in Dammarie-les-Lys, near Melun, Seine et Marne, some 50 kilometres South

---

[22] Dissertation on Eternit: *Evolution Technique et Sociale'1922-1923,* published 1995 on the occasion of a large exhibit reporting the sight of former executives on the history of their company.

[23] Odette Hardy-Hemery : *Eternit et l'Amiante 1922-2000.* Presses Universitaires du Septentrion.

[24] S.V.K. for: Scheider & Van Kerkhove.

of Paris. The idea was to manufacture concrete pipes in order not to let the door opened to his competitor *Bonna* just acquired by the powerful French C.G. E[25].

From 1929, PAM[26], king of the pipe industry since the last quarter of the 19th century, feared the arrival of the first asbestos-cement pipes, imported from Italy by Eternit and competing on his almost exclusive market. Indeed, Eternit France, 'very dynamic under the impetus of *Joseph Cuvelier*, his founder and manager', undertook to manufacture pipes, at first for the building industry. Soon after he began to be interested in the water conveyance market, considered as much more noble. In this purpose, as soon as 1928 Eternit purchased a Swiss licence for the manufacturing of pipes, a second one in Italy for high pressure pipes and a third one for manufacturing top slabs.

01-010.OUralithe advertisement. Olivier Delas

In 1928-1929, the company also took control of one of its three main competitors: Ouralithe from Toulouse, *South-west of France,* acquired from Uralita.

[25] C.G.E.: Compagnie Générale d'Electricité, one of the few big international French groups at the time. Accidentally, C.G.E. also owned Cofpa, paper and asbestos-cement felts manufacturer, where the writer began to work in 1963.

[26] PAM: familiar name given to Société -de- **P**ont-**a**-**M**ousson by his members. By inclination, I will carry on using this name all along the book and you will come across this short form quite often.

In spite of significant investments, Ouralithe was closed as soon as 1931, due to its dilapidated machinery.
Besides, Eternit got associated with *Fibrociment de Poissy* which was the oldest and most important company in France as far as asbestos-cement was involved, counting then almost 400 workers. Eternit, backed by its position, signed an agreement with PAM. This agreement granted Eternit some 70% of the flat products market in which it was specialized. 'In exchange, it would limit its sales of pipes to the building industry'[27].
In 1926, *Uralita, Spanish Cataluña,* probably thanks to its available cash resulting from the sales of *Ouralithe*, acquired a Mazza licence and began to manufacture its first asbestos-cement pipes.
The ambition of Eternit remained obvious. In the same year 1926, it created *Dimatit* in Morocco near Casablanca[28] and opened a workshop for mouldings in its French Prouvy factory. One of the strengths of Eternit was the already mentioned SAIAC, -association of all the Eternit Companies, eight of them already in Europe-exchanging their technical knowledge, which helped

08-011. Eternit Thiant. Olivier Delas

[27] Odette HARDY-HÉMERY (University Charles-de-Gaulle/Lille 3, IFRÉSI – IRHIS, CNRS) **XIV** International Economic History Congress, Helsinki 2006 SESSION 47
[28] Odette Hardy Hémery, Eternit et l'Amiante 1922-2000.

them to save time on their working hard competitors, who were searching solutions each one on its side.

In 1930, twenty-four countries producing asbestos-cement were already listed worldwide. In the same year also appeared the three lingual *International Asbestos-Cement Review* published in German, French and English. Helped by this review, Eternit seduced architects and undertakers, offering them study trips in big European capital cities, proving if still needed, its ability to develop not only its manufacturing process, but also its marketing ability.

Uralita, still tightly linked to Eternit, began to produce asbestos-cement window-boxes and water tanks, which later on knew a wide success in so-called 'developing countries' where water reds remained stuttering.

Also, in 1930 in Japan, the *Tokyo Gas Company* purchased the right to sell asbestos-cement pipes and accessories from Eternit Italy for the equivalent of some five million Euros. This operation permitted the birth in February 1931 of the *Japan Eternit* which started to produce asbestos-cement pipes in 1932. Source: IBAS.

At that time, *Mr Bindschedler,* managing director of *Société Française de l'Everite,* which then counted only one factory in *Bassens, Gironde* got in touch with PAM and suggested to let them have control of his business. PAM did not refuse, but studied with Eternit the future of asbestos-cement pipes for drainage and pressure as well[29]. Eventually they came to an agreement in 1932. PAM, wishing to have a tool against Eternit, accepted the offer of *Mr Bindschedler* and took control of Everite on February 9th, 1933. PAM then joined an asbestos-cement pipe factory to its concrete pipe factory of *Dammarie-les-Lys*, some 50 kilometres south of Paris. It also purchased an asbestos-cement pipe licence to *Dalmine*, another Italian machine-inventor and constructor.

In 1932, appeared a regulation in the United States regarding working conditions when using asbestos. The same year Mexico gave birth to its first asbestos-cement company.

[29] Let us remind the reader that *pressure* reverts to pipes used in delivery of clean water and *drainage* to those used to take away black or rain waters.

In Spain the *Roviralta family* sold his *Cerdanyola* factory to the French Eternit[30]in 1933.
In 1935 in Chile, probably sponsored by the Swiss Eternit, appeared *Pizarreño,* which, along years, became the main producer in the country.
In 1937, Eternit in *Prouvy, Northern France,* created a social service including an *Inner Mutual Fund* and a *Christmas Giving Event*. An Italian asbestos-cement manufacturer created a branch near Sao-Paulo, Brazil.
In the same years, Brits and Canadians enhanced the toxicity of asbestos. At the same time in *Cerdanyola,* worker unions of Eternit rebelled and also participated[31] to the famous and painful civilian war of Spain.
In 1938, PAM united Everite and Economit in: *Sociétés Réunies Everite Situbé*[32].
In 1939, when WWII began:
- At the Swiss National Exhibition, Eternit had its own pavilion.
-Eternit France, once for all, took full control *of Société du Fibrociment et des Revêtements Elo* from Poissy, which was going to become the main asbestos-cement manufacturer in France, already managing some 1,500 concessionaires in the country.
- From the previous year, the French government, forecasting the problems to come, had suggested Eternit to consider a possible fall-back southward, which duly enhance the importance granted to this industry by the French authorities[33]. Eternit consequently had purchased a plot in *Vitry-en-Charolais near Mâcon, around 70 kilometres north of Lyon*. The manufacture, foreseen on the site began to operate several years later. At the time, Eternit France already owned seven sheet machines, shared between Prouvy and Poissy, as well as two machines for 3-meters pipes and one for 4-meters pipes.
When WWII started, the French government insisted on Eternit to realize the project of Vitry-en-Charolais. The result was delayed till March 1944, i.e. just three months

---

30 *Miguel Sánchez Ecología Política n° 37 Salud y medio ambiente* (PP. 105-109).
31 Newspaper article: *La fàbrica d'Uralita (1907-1997.* El Nou, Sabadell, 31-7-2000, p. 13.
32 Most of the information regarding the history of PAM's, *(later on Saint-Gobain)* pipe manufacturers, arise from the memory and private archives of Alain Sabouraud, former manager of Everite R&D department.
33 Exhibition catalogue of Theophile Jouglet Museum in Anzin (North) 1995, due to remind the numerous industries of the area.

before the landing of the Allied in Normandy. The big pipe machine was then operated by non-mobilizable workers transferred from the factories of Thiant and Poissy. Still in 1940, *SMA, Société Minière de l'Amiante*[34] was founded in Paris. The purpose was to make use of the Corsican mine of Canari, known since the second half of the 19th century, when Napoleon III ordered an inventory of the nation's natural richness... The financial resources were said to originate for 65% from Eternit and the balance from Belgium, Switzerland and even Spain, which is quite likely considering the close links between the different Eternit. The history of the mine of Canari in itself justifies a large book. The task was duly performed by Guy Meria who published *l'Aventure Industrielle de l'Amiante en Corse*[35].The weak production of asbestos in Europe during the war was mostly reserved to military needs. Same as during WWI, asbestos-cement was used to construct light sheds in a very short time, for instance light garages for the vehicles of the German army. War did not stop business.

In 1941, Eternit Switzerland[36] created three factories in Colombia. Let us note that such decision might be linked to the fact that this Latin-American country owns an asbestos chrysotile mine, grade 6/5/4, in open sky at *Yarumal, north of Medellin.*

The shortage of raw material made Eternit think to salvage asbestos from waste and to get interested in the Corsican mine as we could see. They even thought to create a new factory in Morocco. They bought a plot and created a company in that purpose, but circumstances delayed the project...

During all those years of war, supplies of Asbest

01-012 23% price rise. Olivier Delas

[34] Mineral Asbestos Company.

[35] *L'Aventure Industrielle de l'Amiante en Corse*, by Guy Meria. Publisher: Alain Piazzola, Ajaccio, 2003. Almost impossible to find.

[36] Thanks to its legendary neutrality, Switzerland does not participate to the war.

and cement as well, were very difficult in France, same as for most raw materials. Substitutes were searched and sometimes found, with more or less luck: pulp extracted from straw, potatoes tops[37] or even rush from the *Garonne River,* as the factory was located on the banks of the *Gironde* estuary. The lack of raw materials helped to accept price rises which would certainly be refused nowadays.

Same as for most activities, the end of WWII in 1945, was a turning point for the asbestos-cement industry. But let us take a few moments to learn more about asbestos, this promising fibre which turned into a misfortune bearer. Let us also listen to what we are told by *Yves Ferry*, former worker at *Fibrociment de Poissy*, reported by Olivier Delas[38].

"I left school at the age of 13 and I was taken on by Fibro Ciment Elo in Poissy. My first job was in the general warehouse, directed by Mr Bidault. I worked eight hours a day and my working time would start on my bike. I had to get ten litres of fresh milk at the Morizet farmhouse, located in Picquenard. I loaded this valuable good in two five litre milk cans, on each side of the handlebar. Then I brought them to Madame Sinot at the canteen room. She would hurry to boil the valued beverage and I'd take it to workers who were greatly at risk, with asbestos dust. My day went on either on errands out of the factory, either distributing mail across workshops. In any case I got back to my bike."

---

37 Odette Hardy-Hemery, Eternit et l'amiante 1922-2000.

38 Extracted from : *Les Etablissements Industriels du Fibrociment à Poissy,* par Olivier Delas. Chrono Poissy, October 2013.

II Asbestos

"A miracle of heat, pressure and time"

Steve Korris[39]

Roman Empire

Nobles of the Roman Empire impressed dinner guests by throwing asbestos napkins into fireplaces and bringing the napkins out white and whole. Roman historians noted that some slaves coughed and died young. They had been weaving asbestos napkins.

Charlemagne

We heard that Emperor Charlemagne, who had a certain knowledge of ancient history, would mystify his tough warriors guests by throwing a napkin made of asbestos yarns into a fire, then taking it out unaffected. The emissaries of rival kingdoms, who had seen such an unbelievable phenomenon, would run back to their chiefs and inform them that they should never go to war against such a powerful king.

Marco Polo,

During his famous Asian trip of the thirteenth century, Marco Polo went through a mountain chain where he came across a seam of asbestos. Remembering the myth of the animal that does not burn when thrown into fire, he named it "Salamander". He made the most of the case to testify that a salamander is not an animal as no animal can survive in fire and he gave a very neat description of the product.

[39] Steve Korris, US expert in asbestos, Madison-St. Clair, 2005 June 17th.

Let us see how Steve Korris describes for us "the creation of asbestos":

"Deep within the earth's crust of a remote geologic age, there occurred a tremendous subterranean upheaval. Masses of molten rock spewed upward, torn by the irresistible forces of terrible heat and pressure. Then hot mineral-bearing waters poured in. Slowly, over a long period of time, the character of the rock underwent certain changes.

Millions of years later, another great uplift took place. Again, the hot waters rushed in, drenching the rock with their mineral-carrying solutions. Again, the rock underwent a change. But this time there was a difference. Now there were numerous openings in the rock and the hot waters could penetrate it intimately. In doing so, they dissolved some of the twice-changed rock, becoming saturated solutions. As water cooled, the dissolved rock was precipitated as crystals. Eventually, the cracks and crevices through which the water had circulated became a network of fibrous crystalline veins that ran irregularly throughout the masses of the surrounding rock. So, the geologist theorizes was asbestos created in nature's underground laboratory– a miracle of heat, pressure and time.

Miracle is not too strong a word to describe this remarkable fibre of stone. Its very name is an index of its enduring character for the word asbestos, derived from the ancient Greek, means inconsumable or indestructible. As nearly everyone knows, asbestos cannot burn. It resists heat and the action of most ordinary chemicals. There is no other fibre –natural or man-made –quite like it.

It would appear that nature was in a generous mood when she created asbestos for deposits existing in many countries throughout the world and there are some thirty known varieties. However, only six of these are of economic importance and one type, chrysotile, is so superior to the others for most industrial purposes, that it accounts for about ninety-five per cent of the total world production of all-natural mineral fibres.

Canada is the largest producer of chrysotile asbestos and the fibres which come from the province of Quebec are noted for their high tensile strength, their flexibility and their extreme fineness. Many of these fibres are as strong as some types of steel wire! Yet they are soft and silky and so fine that a single fibre can be seen only with the aid of the most powerful microscope. "

ASBESTOS- Asbestos is a town located in MRC[40] des Sources, Estrie, Quebec, Canada. Inhabitants are Asbestriens.

No, you are not dreaming. Indeed, Asbestos is the name of a Canadian town, the history of which I am going to tell you:

In 1879, about sixty-five miles from Montreal and Quebec, children brought home a strange stone. Their parents walked with tools to where it had been discovered in order to remove some more stones. It soon appeared that it was the same type of stone as exposed in the nearby town of Thetford Mines. It is greenish and crossed by a "white-grey" strip which can be removed when scrubbed with the nail .

02-001 Asbestos stone. Collected on the former pit of Asbestos City.

The obtained fibres are soft when you touch them: Asbestos!

This discovery gave birth to a fabulous history such as only the huge needs of modern industry can generate.[41]

At the time, the place was nothing but a hamlet which was soon known as "the Mine". Operating started slowly and with plenty of problems. In 1881, the first classroom was opened in a private house. The following year appeared the first shop for ordinary supply, originating the actual birth of the village. Then the mine counted some 550 men and the school greeted 96 pupils, with only one female teacher. Until

[40] MRC means: County local council. MRC des Sources gathers 7 municipalities: Asbestos, Danville, Saint-Adrien, Saint-Camille, Saint-Georges de Windsor, Ham-Sud and Wotton.

[41] Most information about the history of the town are due to Brother Fabian, author of "Asbestos" who, during some forty years at the time of its growth, plaid a major role in Education.

1909, there was no town-hall and the 7 councillors would gather in the house of one of them. In 1895, *Asbestos & Asbestic Corporation* was created with British funds. Thanks to the acquisition of grinding machines, weekly production came-up to 175 metric tons[42].

In 1898, at 12.30 on June 17th, an explosion killed three men. On Marsh 22nd 1900, most of the company was destroyed by fire. In 1901, production resumed and reached 210 metric tons a week.

In 1907/1909, the school counted 124 boys, 103 girls and 4 teachers.

In 1910, an aqueduct and sewers as well as a fire department were planned.

Soon after, the vicar of the only existing parish was in charge of the newly settled phone line.

On January 6th 1911, a new explosion succeeded in the mine. A stone was thrown quite far. It rebounded and killed a six years old girl, daughter of the mayor.

02-002 Mine of Asbestos in the first years.« Asbestos, Filons d'Histoire ».

During WW I, though part of the workers were called-up to defend Europe, the mine pursued its growth and turned to be one of the biggest asbestos mines in the world. In 1916, Asbestos & Asbestic Corporation became "The Manville Asbestos Company".

In October 1918, an influenza epidemic killed 27 kids and 72 young adults, plunging the town in mourning.

[42] One metric ton =2.205 pounds =1.093 short US ton.

II Asbestos

In 1920, the *National Union of Asbestos Mines* was created. It was the first Union in Quebec. One of its main tasks consisted in fighting against Anglo-Saxons, considered as too invasive in the province.

In 1924, the company in charge of the mine again changed its name and turned into "*Manufacture Canadienne Johns Manville* ". It also started to produce asbestos-cement items.

In 1929, appeared the first complaints due to lung diseases resulting from asbestos. At the same time, the company purchased at very high prices, all the properties of the first village in order to increase its operating field. Of course, such policy generated a "boom" on the building industry giving birth to a new urban site.

Then the big depression of the nineteen-thirties arrived. In order to provide occupation to idle youth, hockey teams were created. Yet, as soon as 1934, the pit was newly growing and reached more than 900 meters (almost 3000 inches) in diameter and 61 meters (200 feet) in depth. A half static derrick drilling machine was built on site. Drilling rods were 1.40 meters (4feett 7invhrd) in diameter and 12 meters (almost 40 inches) long. They were manufactured in Canada, with special hardened steel able to withstand the 1200 per minute strokes of the mechanical drill hammer.

In 1930, the company was offering a valuable gold watch to fifteen miners working there for 25 years.

In 1935, a real town hall was built, while the mine was reaching 1070 people. At the same time the mine built its own hospital equipped with the most modern devices for control and research. Yet, a week-long strike enabled the workers to obtain satisfaction to most of their claims. Archives of the time made notes of devastating effects due to alcohol, and managers tried to limit this curse by every possible mean.

In 1940, cheap accommodation began to be built. It was an innovation in the *Belle Province*[43]. Soon after the beginning of WW II a corvette of the Canadian Navy was named "*Asbestos*". The inhabitants of the town created a cloth made of asbestos yarn and offered it to the crew. The corvette performed seven escort trips from Newfoundland to Ireland and was decommissioned in Sorel, Quebec, Canada, in 1945.

[43] Belle Province is the name usually given to Quebec region by French speaking Canadians and French people, could be translated by "Smart County" or "Pretty Province".

Sold to the Dominican Republic, she sank on her way to her host country. Her bell was given to the town, who keeps hold of it in the new boardroom of the city council.
In 1947, a cenotaph was erected in memory of the sons of the town killed during both world wars. They were eight during the first one and thirteen during the second one.
From 1948, the town started selling its streets, for one Canadian dollar a piece to the company. Then Asbestos City borrowed in order to rebuild in other places. The system lasted till 1969, generating a quite fabulous public debt.
In 1949, on February 14th at midnight sharp, the workers from four mines around Asbestos City, all of them owned by US or Anglo-Canadian companies, undertook a large strike, which lasted till June 30th.
Most miners were of French language. The strike was illegal, yet it came to be one of the longest with more violence in the history of the country.

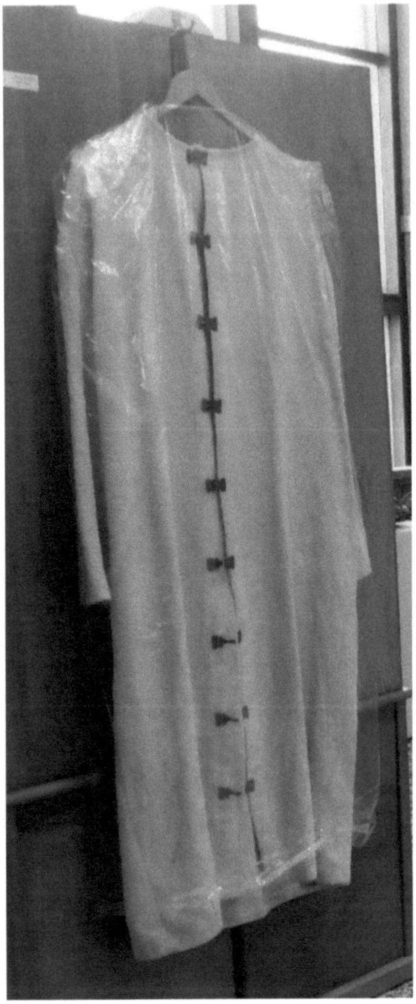

02-003 asbestos yarns Cloth made by the inhabitants of Asbestos City, offered to the crew of the corvette of the same name. Asbestos Library.

Unions were claiming on several points:

-elimination of asbestos dust both inside and outside the factories,
-pay rise by five cents per hour,
-five cents more per hour by night,
-set up of a social sanitary fund run by Unions,

-implementation of the Rand Formula[44]
-double salary on Sundays and public holiday.
Every above claim was totally rejected by the managements of the companies.
The government of the *Belle Province* supported the companies and ordered police to protect the mines. Usually, close to the Catholic Church, it broke with this one latter on, as part of his members supported the strikers. Population and news-papers were also rather in favour of the strike action. Pierre Eliot Trudeau, future Prime Minister, journalist at the time, covered the event and reported it in a rather favourable form.
After six weeks, Johns-Manville hired strike-breakers and some of the workers passed the strike-pickets. The strike became violent and more police force was sent on site. Genuine outbreaks of fight occurred between miners and police; hundreds of strikers were put under arrest. On March 14th, part of the private railroad line of Johns-Manville was destroyed by an explosion. On the 16th a jeep belonging to the company was turned upside-down and one of the passengers was hurt. On the 18th, a senior executive of the company was pulled out of home and actually beaten-up.
On March 5th, the Archbishop of Quebec preached in favour of unions, suggesting his flocks to financially support the strikers. The prime minister invited the church to move the Archbishop to Vancouver. The archbishop resisted and ended up as a chaplain in a British-Colombia hospital... On May 5th, strikers attempted to close the mine in Asbestos-City and to block the roads leading to town. Attempts of the police to remove the barricades by force failed, but strikers withdrew when police threatened to fire live ammunition. The next day, Riot Act[45] was claimed and massive arrests were carried out. Detained strikers were beaten up and their leaders were brutalized.
After these arrests, unions decided to negotiate. The new Archbishop of Quebec was named mediator. In June, strikers accepted to come back to work with very few advances. They just benefited of a light raise in their wages and lots of them never got their job back.

---

[44] Rand Formula: law forbidding that anyone who joined a trade union can withdraw from it as soon as wage profits have been obtained, in order not to pay his subscription.
[45] Riot Act: law authorizing authorities to disperse demonstrators when considered as dangerous for general interest

The strike would certainly not have lasted so long without moral and financial support from a part of the population. It came to be a very important component of the development of Quebec. Eliot Trudeau, future Prime Minister, published a book, *"The Asbestos Strike"*, presenting this event as being at the origin of modern Quebec.
In 1953, the church of Asbestos City commemorated the tenth anniversary of the Christian Workers League, which was the original birth of Quebec Trade Unions.
In 1956, a "very great mill", 14 stages high, was built and equipped with the most modern machinery. On the opportunity, the Archbishop of Sherbrooke celebrated a solemn blessing in the presence of Maurice Duplessis, Prime Minister of Quebec. This new mill could process 625 000 tons of asbestos yearly, making thus Johns Manville the biggest asbestos producer in America.
In 1957, mail was at last delivered at home. Some 1300 letterboxes were removed from the post office.
In 1959, the town allocated a 2000 C$ grant to a young girl, good at music, to help her to study music in Paris.
In 1960, the Catholic National Union celebrated its 25th anniversary. She congratulated herself for her large involvement in improving worker's life during her 25 years of existence.
On May 20th,1962, a gas link caused a huge blast destroying two houses and killing five children and three adults.
During 1970, Johns Manville Corporation was newly planning to extend its operating field. In order to avoid collapsing risks which were more obvious every day, the town council ordered to close several shops. At the same time, the town was planning to build low price homes in the long run, while the company claimed to have enough trailers available to face any urgency.
In the very beginning of 1971, the town had to face unkind remarks from Canadian and US media about creaking and clefts on the fringe of subsidence. On 21st January 1971, the famous store *Continental* partly fell down, causing damage to the aqueduct and to sewers. On the 22nd, the *Fina* service-station collapsed in turn and nearby buildings were cracking everywhere. On the 24th, shops moved and settled in makeshift buildings, while engineers closely watched ground motions. Constructions of low-cost homes were accelerated. Altogether 151 private accommodations were

built in a very short time, giving birth to a new estate. The same year 1971, Johns Manville Corporation acquired new trucks, some fully loaded, weighing up to 200 metric tons. Some data can give us idea of those monsters which we now use to see on TV when a big public work or an open mine is involved.

Length: 13 meters, (43ft)
Width: 6.20 meters, (20'4i")
Height (dumpster up): 12.40 (40'8")
Engine oil: 757 litres, (200 US gal)
Cooling: 416 litres, (110 US gal)
Fuel oil: 3000 litres, (792 US gal)

02-004 Giant trucks of Johns-Manville in 1976.

Those gigantic trucks, as well as the newly arrived machines, reduced the number of workers. Consequently, from 1973 on, population began to reduce, though production still stood at a very high level.
At the beginning of 1975, town authorities started to fight against the managers of the mine. Then a new mudslide came that revived the still not cured previous one, which had occurred four years earlier.
Though activity was still intense, one can say that happiness was gone. The future was no longer certain, the town continued to lose inhabitants in spite of the efforts made by the elected members.

At the end of the 20th century, the town of Asbestos was ready to celebrate its first hundred years. It still believed in a bright future, thanks to the foreseen use of the gigantic waste buried in its soil. Magnesium was in a very great need worldwide and it could be extracted from the above-mentioned waste. Unfortunately for the town, ecology and very cheap Chinese magnesium coming together soon killed the beautiful dream around 2012[46].

02-006 Asbestos Quebec Canada Former pit of the mine, now a lake. Photograph by Emilie Roulland

[46] To the reader interested in a detailed history of the town, we recommend *"Asbestos 2eme edition revue et corrigée"* by Frère Fabian, and *"ASBESTOS Filon d'Histoire 1889-1999"* published by 3 historians, Rejean Lampron, Marc Cantin and Elise Grimaud in 1999, at Transcontinental on the occasion of the town's centenary. Unfortunately, we do not know if these books have ever been published in English.

This same year 2012, after more than one hundred years during which more than sixty million metric tons of asbestos had been produced in the country, Canada definitively banned the use of the fibre.

One will note with interest that the population of Asbestos tends to minimize the toxicity of the fibre, origin of its name, and seems to show a kind of nostalgy of the old days...*(Information pointed out by the author's grand-daughter who visited the town, the site of the ancient mine, and talked with several inhabitants).*

Let us leave the subject on a more optimistic note and have a look at the beautiful blue colour of the lake appearing on the site of the former open sky pit.

But Canada is not the only country owning a town named Asbestos. Russia, still nowadays very big producer and exporter of asbestos, also has its Asbestos City on the East side of Urals. The town counts some 70,000 people and seems very flourishing. Telling its history would not bring us much. But we can refer to Nicolas Badiotel who counts in the French newspaper *La Croix* dated April 24th 2017:

"Here, it is quite frequent that just married couples have their happy-wedding-snapshots made on a platform offering a wide view to the great pit of the asbestos mine. Such panorama offers both a pleasant and promising background to the photograph. They have a quite different sight of asbestos fibres".

One tends to think that they are denying reality in a very similar form to that of the people in Asbestos Quebec, Canada.

And what about France? Will you tell me? Yes, in France we had mines, too.

At the end of the 19th century, the French government ordered an inventory of the various mining richness of the country. In Corsica a shepherd of the village of Canari (*north-eastern coast of the island)* had discovered a fibrous rock in the mountains. He brought it to the mayor of the village who sent it to Paris. The rock was identified as asbestos and its presence was registered in due place.

In 1928, in Chateau-Queyras, Hautes Alpes, South East of France, now Chateau-Ville-Vieille, an asbestos deposit was discovered but remained almost unexploited until 1940. In 1942, it was requisitioned. Activity was developed especially from 1947 when samples sent to Canada for analysis proved that the quality was good. Though the deposit could be operated in open sky, the altitude of almost 2,500 meters

(8200ft) required enormous works of infrastructure, such as a 12 kilometres (7.5 miles) long dizzying mountain road, a cable car to carry ore from the mine to the mill located 1000 meters (1,100 yards) lower, and an electric-power plant to supply energy to the whole facility[47]. In spite of those costly investments, the site had a rather short life...

During WW II, the lack of supply of imported raw materials forced industrials and governments to search every possible resource on the mother land.

That is why, the Corsican mine of Canari already mentioned, but lightly operated, all of a sudden became very attractive. Eternit, Everitube-Sitube, as well as other companies from France, Belgium and even Germany tried to join forces and created SMA, *Société Minière de l'Amiante* (Asbestos Mining Society) in order to manage the mine. But indeed, actual exploitation only started after the war. Very soon it appeared that the asbestos coming from the mine was not of good quality as usually imported from countries such as Canada, Russia or Rhodesia, which were then coming back on the international market. Shareholders refused new investments and the mine was closed as soon as June 1965. The post-result is an outrageous brownfield, as testified by several press articles and documents about this matter. Guy Maria's book, already referred to, is very clear in this regard. We will have the opportunity to revert to later on.

In most countries one can note the slowness from authorities to force industrials to conduct serious actions in order to protect people working in the field, though dangers were long known. Such situation can be explained by the power of lobbies, not only from asbestos-cement industry, but by all those using asbestos such as: automobile, house and shipbuilding industries. Anglo-Saxon countries developed a theory, adopted with excitement in France around 1970, known as "controlled use". This technical solution claims that respecting its advocated norms, risks are indeed inexistent. Quarrels between representatives of industrials, state departments, doctors, experts and lawyers are endless. Victims apply to courts, patients die and families cry, but nothing seems to work. Years elapse and situation gets worse.

---

[47] All the above details, as well as those from the following paragraph are taken out of "l'Aventure Industrielle de l'Amiante en Corse" Industrial Adventure of Asbestos in Corsica by Guy Meria, at Alain Piazzola, 2003.

France finally bans asbestos in January 1997. We will revert in details to this point when talking about the fibre-cement crisis.

In most countries, industrials blackmailed members of parliament, ministers and heads of states, with the endlessly repeated "job-cuts" menace. This is one of the reasons that delayed the ban in spite of the conclusive evidence of asbestos toxicity in the long term, and even in the very long term. On the other hand, it must be noted that the numerous factory managers that I met and interviewed when preparing the present book, still have the sincere feeling that they observed the current regulation of the country where they were working.

As to know if the first responsibility belongs to industrials or to the states, do not expect me to decide. I do not feel to be able to, nor to be allowed to. An interesting document from Robert L. Virta[48], published by "USGS science for a changing world", offers a wide explanation about asbestos in the world between 1900 and 2003.

AFP[49] and the French evening newspaper Le Monde, report: *"On June 12th 2017, the Public Prosecutor's Department of Paris states in his petition: The diagnostic of an illness due to asbestos proves the poisoning, but does not allow to know neither date of exposure, nor date of contamination".*

In the same month of June 2017, the Court of Justice of Turin, Italia, delayed to December the second proceedings of Stephan Schmidheiny...In 2003, the first proceedings had condemned him to an 18 years jail sentence for his personal responsibility in an environmental disaster linked to the Italian fibre-cement factories belonging to "Eternit Spa Genoa". The Supreme Court of Italy had acquitted him in December 2014.

A new alarm was introduced by the French media "Conso Globe- La Planète Vivante", who wrote in June 2017:

"***World asbestos production is newly increasing!***

"Nowadays, countries with a high rate of growth develop their dawning economy by intensifying their asbestos industry, either through mine exploitation or factory transformation: they are searching new

[48] Robert L. Virta is a US scientist and writer who studied asbestos and published very interesting books and articles about the material.

[49] AFP : Agence France Presse.

markets linked to this raw material. The result is that the total volume of the world production of asbestos, which was reducing due to a lower consumption in industrialized countries, is meeting a new rise since a few years, in spite of the clear international scientific consensus as to the hazards involved by this product".

Mineralogy	Type	Chemical composition	Properties	Main deposits
Serpentine	Chrysotile (white Asbestos)	Hydrated magnesium silicate	Fibre with high flexibility and strength. Attacked by acids.	Canada (Quebec) Brasil Zimbabwe Russia (Ural)
	Pierolite	Idem chrysotile	Lack of flexibility.	Few (Sweden)
Amphibole	Anthophyllite	Hydrated iron and magnesium silicate	Breakable	USA (Massachusetts)
(Rhombic type)	Tremolite	"	More brittle, less resistant to heat. Resistant to acids.	Italy, Africa, The Balkans
Amphibole	Actinolite	Hydrated iron, calcium and magnesium silicate.	Brittle	North America
(Monocyclic type)	Crocidolite (blue asbestos)	Hydrated sodium and iron silicate.	Blue, flexible.	Cape
	Amosite	Hydrated iron silicate.	Less heat-resistant than chrysotile. Acid resistant. Long fibres, can be woven.	Transvaal

III Manufacturing process and products

> *"On nights of victories, one imagines that never, never, never more defeat will happen, and on nights of defeat, one figures that never, never, never more victories will happen. But when one is an old soldier, Mrs. Jane, one knows the truth* [...]. *I have seen so many defeats happening after victories and I also have seen so many victories happening after defeats that I never believe that it is over. "*
>
> Charles Peguy[50]

Originally the newly invented material was named "asbestos-cement", by inventors and first developers, which sounded very logical as the two main components were asbestos and cement. But soon, the name "fibre-cement" or "fibrocement" became the reference name. It seems that it was borrowed from the French company *"Les Etablissements Industriels du FIBROCIMENT de Poissy"*, but that is just a supposition, in any case most people remembered, and still remember, this designation. In France, some companies did not wish to use the name considering that it belonged to a competitor.

The manufacturing process for flat sheets, slates, corrugated sheets moulding and pipes have all a common point: the "soup" or "slurry", i.e. the prepared mixture of asbestos and cement is diluted with a rate of more or less one hundred grams of dry material to one thousand grams of water (10%), when arriving to the sieves of the Hatschek machine. Concentration and adjuvant may vary according to supply of

[50] Charles Peguy. 1873-1914, famous French writer and poet, killed on the front in the first days of WW I. *"Jeanne d'Arc-Les Batailles 1897 "*, gives this lesson to John of Arc through one of his old comrades.

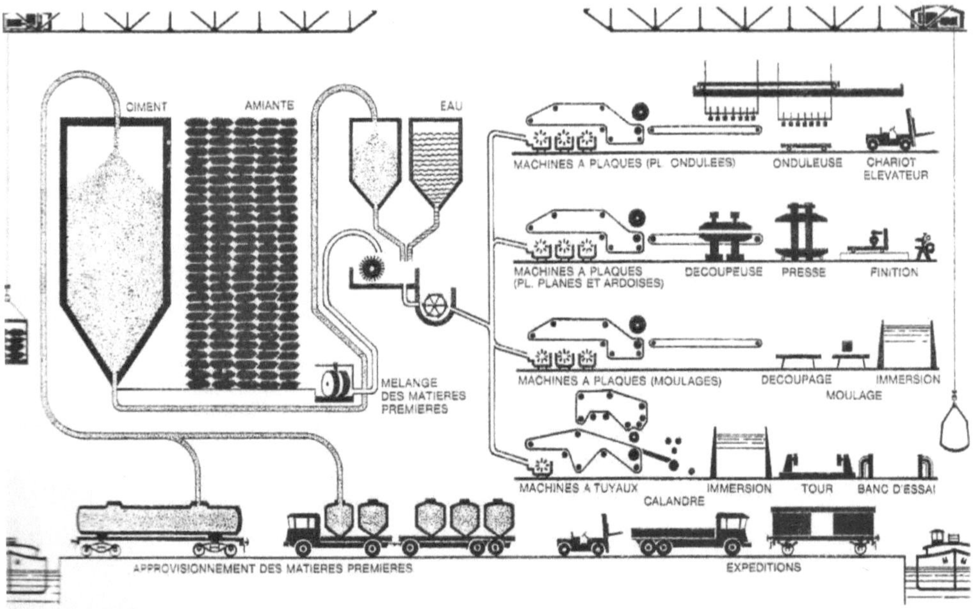

03-001 Asbestos-cement machines scheme out of "Eternit et l'Amiante 1926/2000" by Odette-Hardy-Hémery. Septentrion editor.

the moment. Costs, political situation and level of quality desired may also generate different choices. Regarding cement quality, Portland cement is used almost every time and everywhere, due to its wide availability. This Portland cement also grants a fast hardening - not a quick setting- phenomenon which is considered as very important in the case of mouldings and in moving steel templates, subject on which we will revert later on. Moreover, it also has the advantage of supporting various curing conditions, a matter which we will equally have the opportunity to revert to.

Asbestos, originating from the different mines scattered over the world, had already been extracted from the rocks in which it was confined, most time right on the mine site. It used to arrive to the factories in jute bags, later on in polypropylene[51] bags, often pierced along their large journey by the hooks of those who handled them with

[51] In the 1970, a rule obliged supplier to use polypropylene bags. Russian suppliers were difficult to convince. (Memory of an ancient member of the French Everite).

more or less care. *(Some deliveries were made loose in barges or containers, then re-handled, without much care, by the workers in charge of moving them to the local warehouse)*.

Around 1970, a system was developed which permitted to open the bags in a closed place, asbestos could then get out and go directly into the preparing blend. Empty bags would fall on the other side into a mechanical device which would chop them into confetti. These ones were then sent to the blend without transit in the open air. Before being incorporated in the paste, asbestos had first to be "defibrated" or "grinded", which means that the naturally aggregated fibres had to be separated one from the other, while taking care to maintain their entire length. Only high-quality asbestos offered a financially acceptable productivity. This implied and justified research travels to producing countries from the part of industrials using it. Such a situation sometimes generated pacts between competitors, in order to obtain interesting purchase prices, without loss in quality.
A great many devices were conceived by the people in charge of asbestos mines, or by those of the manufacturing firms, regarding the treatment of asbestos fibres. Treatment was a final determiner as to process and quality of the end product. The oldest one was the so called "Hollander", borrowed from the paper industry. The purpose of the operation was to minimize the small blocks or small logs, resulting from the conglomerate of a great number of fibres constituting raw asbestos. Here again, a great care had to be applied in order not to break the fibres and to maintain their original length, which was a main factor for the endurance of the end product. But the most commonly used device was the "edge-runner", derivative from the traditional mill grinder. This "edge-runner treatment" could be performed on dry asbestos or on previously humidified asbestos.
Once the grinding was operated, though maintaining a certain cohesion, fibres adhered only lightly one to the other. Then it was necessary to separate them. This new operation called "disintegration" was performed with "fans" turning at high speed inside airproof metallic cylinders where the airflow violently threw the fibres against the inner sides, though without breaking them. Then, they were sent to a box in which they were mixed with Portland cement in duly defined percentage. Every above operation was carefully checked through samples which were analysed in the own laboratories of the manufactures. These laboratories used to be located very

near from the fibre-cement machines, so that results were known as soon as possible and corrections be immediately applied when needed.
Among controls performed in laboratories, one was very important. It consisted in determining the actual characteristics of the received asbestos. One of the most important related to their theoretical value (TV). This basic information was required to find out the exact quantity of asbestos to blend with the cement in order to grant the final quality of the product to manufacture.
Mixing with very big quantities of water was performed with a process very similar to that used by the paper industry or rather to that of the cardboard industry. It could slightly vary according to the products to make and to raw materials available, but up to this stage, one can say that processes were very near one from the other. The general diagram shown in the beginning of the present chapter *(picture 03-001)* can be considered as a good summary of the most frequent types of manufacturing described in the following pages. Each with its proper required machinery.

Sheets:

This is where the idea of Ludwig Hatschek revealed brilliant. Mixing asbestos and cement in huge quantities of water, thanks to the cardboard machine available in the family enterprise producing beer coasters, he invented a process which the whole world soon started to use in multiple forms.

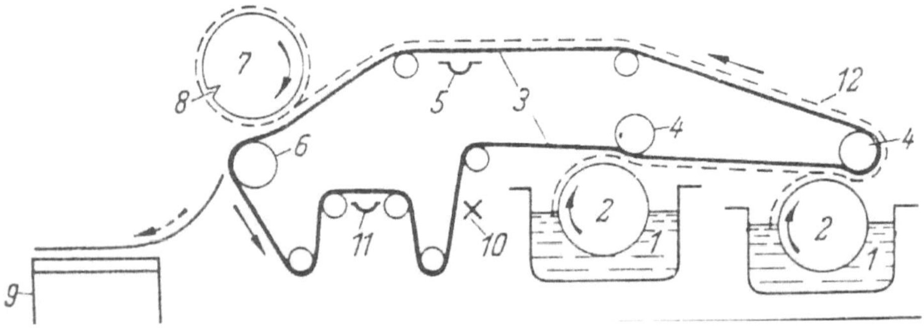

03.002 Hatschek sheet machine , conventional scheme

The paste, often called "solution" or "slurry" very fluid, was sent to the **vats** (1) in which **cylindric sieves** (2), were turning a *thin layer of paste* deposited on their surface and a large part of the water was filtered. Due to capillarity, this thin layer adhered to an **endless belt: the felt** (3). The felt was **turning endlessly**, and **conveyed the paste** (12) to the **making roll** (7), on its way it passed over more or less numerous and powerful **suction boxes** (5) which intensely participated to reduce the percentage of water remaining in the paste. When the paste layer reached the also turning making roll (the surface of which was smoother than that of the felt), it adhered in turn to the roll thanks to the difference of capillarity. The single layer **winded on the roll,** and was pressed by the action of the **motorized bottom roll** (6). The single layers stuck one to the other under the pressure. After 4, 5 or 6 turns of the roll, the desired thickness was obtained, a cutting device made the forming sheet fall down on a **conveyor belt** (9) which carried it to the next step. The felt followed its way going back towards the sieves. In between it passed over a cleaning system, made of rolls, **one or two wipers** (10) and a **suction box** (11) settled to extract water out of the felt, so that it will not be saturated when picking up new layers from the sieves. Asbestos and cement loaded waters were driven back into the circuit through tanks, pipes and adapted flow gates.

Over the years, companies proceeding with the system improved the machines with more round sieves and more suction boxes, better control of dryness, mechanical cut instead of manual cut and all parameters allowing to accelerate the whole process. The goal always was to improve both, quality and a number of square yards produced. Not only was speed increased, but the forming cylinder saw its diameter enlarged so that sheets could be longer[52]. In some cases, they came to reach 8 yards in length and even more for some special sheets. Improvements were most im-

[52] Remember that the length of sheet is linked to the diameter of the making roll and the width to the length of the same roll, which in turn corresponds to the nominal width of the machine.

portant in the 1980's when management and control of parameters became computerised, often inspired by those used in the paper industry and sometimes just reproduced from them. Not only did they improve productivity, but they also made qualities steadier. In spite of all that, the initial process remained, and still remains in use in our 21st century, even if asbestos has been replaced by substituted materials.

When falling from the making roll, sheets were ready to receive different alterations according to their future use.

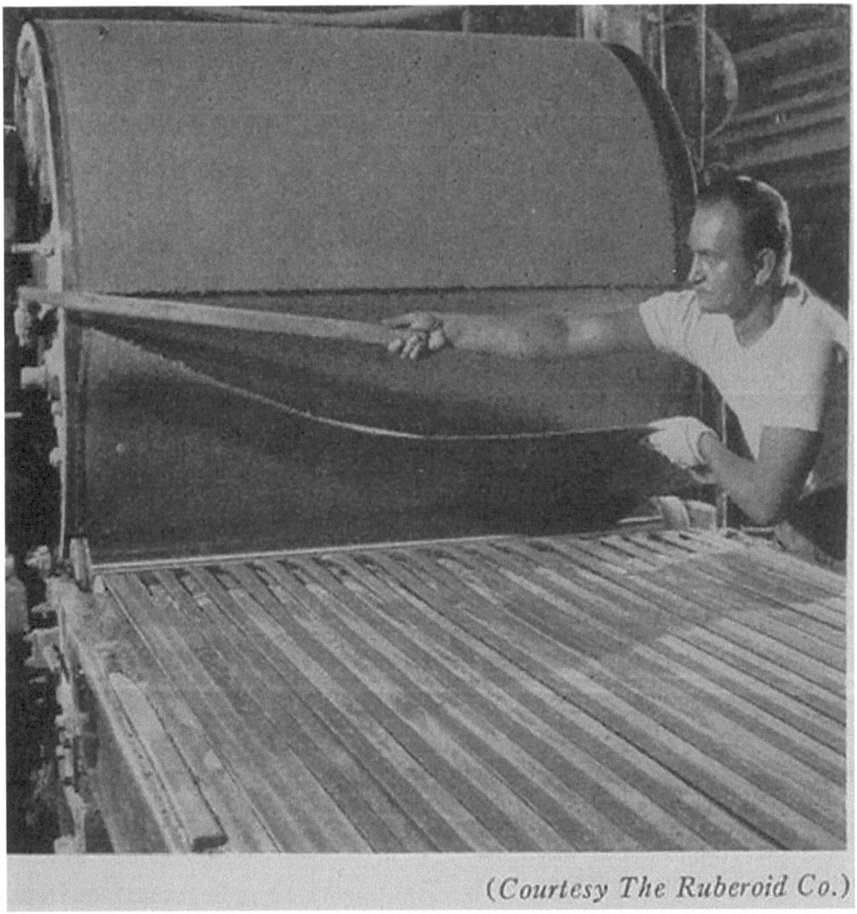

03.003. Hand-cut of a forming sheet. Curtesy of InspectAPedia.

Flat sheets:

Flat sheets were the easiest to produce. Once the edges had been shaped by more or less mechanised rotating discs, later on by high pressure water jets devices, the fresh sheets were piled up, separated one from the other by flat and oiled metal sheets used as separators usually called "templates"[53]. When the sheets were intended to make slates, the pile of sheets and their templates had to pass in a high-pressure press, up to 400 atmospheres, and they lost about 10% of water to reach a minimum porosity, a good mechanical strength and a limited sensitivity to weathering. This last quality was of course basic as their purpose was to cover roofs, or walls in vertical position most time on the side of prevailing winds. The setting of cement required some twenty-four hours. The templates could then be removed and stored for a future use, and sheets could be piled directly one on the other. One still had to wait about four weeks to obtain a complete cure which would make them ready for sales. Binders introduced in the paste, as well as finishing, also varied according to the final use of the sheets.

Some sheets needed to be air-steamed, which required first to "de-stack" the piles and to replace the original templates by others, specially conceived somewhat like grids, in order to let steam pass through. Air-steam consisted in letting these piles of sheets inside the air-steamer, steam saturated up to 170° C (338° F) during 8 to 24 hours. This method permitted to avoid the usual 3 weeks of cure, consequently placing the sheets on the market much faster. This was really true when energy was not yet a problem. Let us recall that, except in wartime, until 1974/75, no one would bother about heating invoices, neither among industrials nor among private consumers.

The process was mostly used in case of sheets to be used as sandwich panels for agricultural buildings or prefabricated class-rooms, plenty of which were built in France in the 1960s/1970s.

[53] The name of these metal separators varies according to countries and even to companies. Among them: **template**, which I will use more than once in the following pages, but also: Steel Sheets, flat steel sheets or flat moulds in the case of flat sheets, corrugated steel sheets or corrugated moulds in the case of corrugated sheets.

The percentage of air-steamed sheets was quite variable from one country to the other. Rather marginal in Europe it was very frequent in North America and Australia. For instance, James Hardie, whom we will talk widely further on, used big quantities of siliceous sand, crushed to a fineness near to that of cement, for one part because it was available nearby his factory, for the other because of its very interesting physical properties in the context of air-steaming which James Hardie was largely using.

Flat sheets and slates:

A problem met with artificial slates is their greyish aspect, due to cement. Asbestos-cement manufacturers have always been fighting to offer artificial slates in a more attractive tone and to make it last for many years. From the first years of the asbestos-cement industry, inorganic pigments were introduced in the paste, mostly ferric oxides, but colours resulted rather pale and soon vanished under the influence of the weather and calcium scrumming. In order to solve such disadvantage, they added a coat of paint which tended to vanish after five or ten years. But during that time, sheets had taken age and the initial mass colouration "would partly compensate" and at the end, the result was quite acceptable. Over years, the main companies held important research hoping to find durable solutions, for instance mixing black carbon in the paste, or improving paint qualities, but they never happened to be perfect.

Around 1970, one of the manufacturers, in view to save costs, chose to use black carbon instead of ferric oxide for mass colouration and added paint on the surface. A few years later, when paint was vanishing and customers were complaining, it was discovered that black carbon did not link with cement. It was eliminated and washed by the continuous changes of hygrometry. It then appeared that the only solution was to repaint the slates on the roofs...During years subcontractors made their wealth with this unexpected market which was falling from the sky... and was costing fortunes to the company. Consequently, managers decided to revert to iron oxide. But still with the idea of saving money, they incorporated it only in the last round sieve, the one supplying the surface layer, yet a small percentage of black carbon was mixed in the mass.

03-004. Roofs and walls covered with fibre-cement slates of different ages.

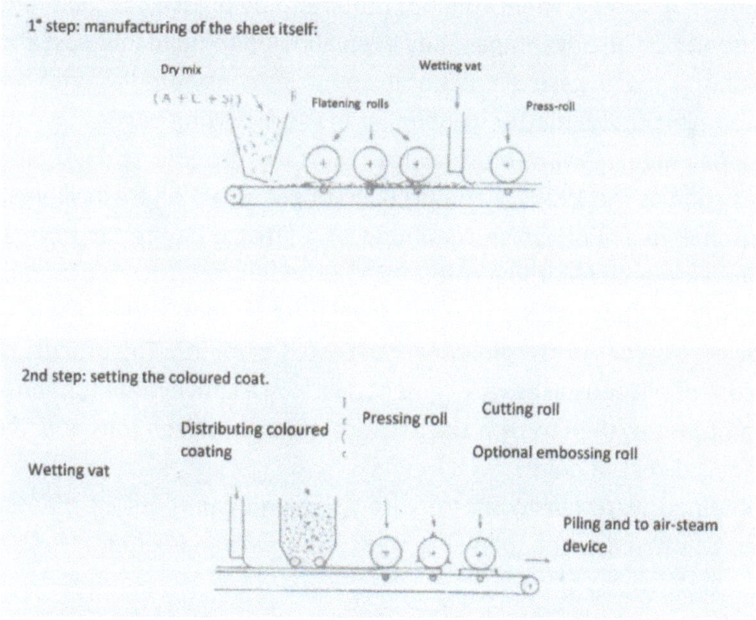

03-005Coloration in two steps. Private archives

Another manufacturer used a different method, called "superficial seeding", which consisted in dusting a colouring blend on the surface of the last layer, still on the felt, during the last round of the forming roll. The one making the surface of the sheet. This technic seemed to give the best results.
Manufacturers of black carbon were not idle. In the 80s/90s, they offered new qualities much stable when mixed with cement. As soon as 1930, in Niederurnen, Switzerland, Eternit had finalised a process permitting to obtain an enamelled surface of the sheets which consisted in passing them through an electric oven. This passage in the oven closed the pores, preventing the devastating scraping effect of snow when melting and sliding on roofs. Such treated sheets were still good looking after ten winters. One can imagine that the idea of a passage through the oven might have helped the birth of a flat product, remarkable by the outstanding quality of its surface: Glasal, which has been one of Eternit's flagships. We will talk about it in details later on. Flat sheets were also used as inside or outside lining for walls. Their isolation capacities were highly praised whenever material used in building construction were insufficient. In that case, sheets did not require to be pressed as their natural porosity appeared to be an advantage. They even allowed to build low-cost constructions in a very short time, in case of crisis or in countries considered as poor ones. Opportunities to revert to the matter will appear in the following pages.
Colouring flat sheets meant more or less the same trouble as darkening slates. At Eternit, as soon as the 1920s, a coloured coating was laid on fresh sheets, just fallen from Hatschek machine, which reminded of a "fresco" quite far from any kind of enamelling. Research regarding colouring was underway in every dynamic company. Let us look for instance at the process developed by John's Manville in Canada, reported by J-P Guerber in his "glossary of asbestos-cement". The process consisted in laying a coat of siliceous-asbestos-cement paste on a conveyor belt, then to humidify it with some water, then to compress several times between rolls and, if needed, to add a coloured coat of paint.
Panels obtained were cut to the required size and came out of the machine stiff enough to be carried to the following steps: trimming and curing.

Many manufacturers also used another type of machine, called "flow-on", especially when making mouldings. Preparation of the paste was similar to the Hatschek machines, but instead of transferring the paste to the felt via round sieves located in vats underneath the felt, paste was just directly laid on the top side of the felt, using a dispensing device, as shown on the following diagram.

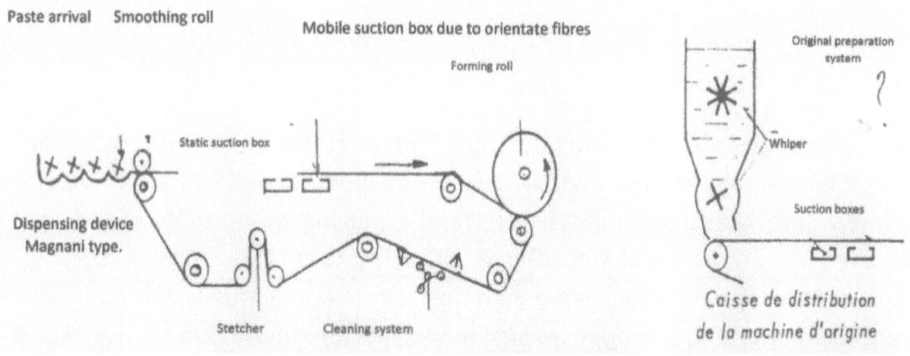

03-006 Flow-on machine diagram

007Modern "Flow-On" machine from unknown source

Bottom right in the picture, one can notice the dispensing device depositing the paste on the felt. On the felt itself, the darker perpendicular strips result from the reduced quantity of water in the paste, due to the action of each suction box right under the felt and linked to pumps by big pipes visible on the left. In the centre of the picture, one can see the making-roll on which the paste layers are winding and sticking one to the other before being cut and separated by the cylinder once the expected thickness is reached.

Corrugated sheets:

Corrugated sheets are probably the most familiar to the general public, especially in France where they knew a very big success all along the 20th century. They are still visible on the roofs of manufacturers, warehouses in towns and in countryside. They are even still present as siding in huge and ugly industrial buildings According to regions or country where they were manufactured, people usually named them simply from the name of the company which made and sold them: Eternit, Everite, Uralita, Duralite etc.

In the first years, they were made of the same size as corrugated metal sheets, which highly participated to their success. Yet, it soon appeared that sizes had to be adapted to use. They were the cheapest of all roofing available at the time when they arrived on the market. They could easily and rapidly be settled, much faster than slates or tiles. Moreover, unlike metal sheets, they did not rust. Single covering of one wave, or even a half one, from one sheet to the next one, was sufficient to ensure waterproofing. The counterpart of above advantages was that sheet fixers had to be very careful and were supposed to always use safety boardwalks.

Among unexpected advantages proper to asbestos-cement corrugated sheets, appeared their capacity to retain condensed waters, due to the hydrophile characteristic of asbestos and to the porosity of the manufactured product. When a cowhouse or sheepfold was roofed with metal sheets and cattle came back in the evening, it generated heat, which caused evaporation naturally concentrated under the roof. At night, when it turned colder, this water dropped on the cattle. When the roof was made of asbestos-cement corrugated sheets due to their constituents, they tended to absorb the water and to let it drop later on, when daylight was back and cattle was newly away.

The need for inexpensive roofs in industrial and agricultural buildings was originally a wide success met by such sheets. The first ones were made in Niederurnen, Switzerland in 1912.

Manufacturing began exactly with the same process as for flat sheets, except that the blend, cement-asbestos, had to be adapted to the specific needs of each type. The best humidity rate required, when sheets came out of the machine, gave rise to research and delicate developments. It mostly had to do with pressure and suction boxes of the Hatschek Machine[54] so that during the following steps, sheets could be easily corrugated without losing their mechanical and waterproof qualities.

When falling from the making roll, sheets were laid on oiled template sheets, no longer flat but duly corrugated in the desired waves, in order to act as mould and also as support during the various motions required before the sheets be ready for use. In the first years, workers had to participate to the moulding. They would manually dispose round poles of the required diameter and length on the sheets. These latter were still flat and flexible, but being laid on the above-mentioned corrugated templates and pressed by the poles, they would take the shape of their metal mould. The first sheets were made with small waves of such a short bending radius, that the paste texture was often damaged during the corrugating process. Moreover, they could not support high pressure, which limited their use overall in areas of intense snow falls. For this reason, they were long considered as a second quality product.

One can note that they proved successful in Great Britain and the Netherlands, as soon as the 1920s, but only from 1932 in Switzerland. Why? Simply because oceanic climate, rather wet and mild, suited them better than alpine climate where snow thawing was very harmful for the first generations of corrugated sheets.

Around 1970, the French manager of a German asbestos-cement manufacture, owned a house in Haute-Savoie, French Alpes. He ordered the roofers to settle corrugated sheets on the frame before placing the tiles. Professionals of the region laughed at him. But the system soon proved effective as soon as snow began to defrost. Water, which usually infiltrated between tiles and stained the ceilings, flew

[54] Let us note that Hatschek Machines were often called "cardboard machines" by their users, due to the original purpose of such machines, used to manufacture cardboard.

down along the hollows of the sheets towards the gutters avoiding the usual spring damage.

03-008. Hand corrugated sheets. Unknown origin

03-009. Very old corrugated sheets. Private archives

At that time, Eternit and Everite developed products designed for this purpose. They met a great success in areas traditionally with roofs made of tiles. Such method is still in use nowadays, even if the modern sheets are no longer made with asbestos as we will see later on.

III Manufacturing process and products

Along years, corrugating technics were progressing. During the 20th century, general mechanisation permitted to maintain acceptable production costs in spite of frequent salary increase in certain periods. This mechanisation, successfully completed by engineers, gave birth to an impressive and perfectly synchronised "ballet of sheets and templates" which, thanks to multiple rubber-suckers in a deafening noise, would let the unaware visitor eyes wide open.

The fact of achieving thicker sheets, up to six millimetres (0 15/64in) and wider bending radius, thus escaping the pale copy of metal sheets was a triggering factor of reliability and commercial success. A greater care in junctions also improved quality and sales results, favouring the interest for the product in many countries.

Adding an entire set of accessories, specially conceived such as ridge tiles or wall and gutter fittings completed the wide variety offered by manufacturers to their customers. From the first time, industrials had insisted on the durability of their corrugated sheets. Actually, many asbestos-cement roofs, even if they let see their old age through adhering moss, are still perfectly waterproof sixty or seventy years after their setting.

03-010 Fisherman's shack covered with asbestos-sheets near Arcachon, Gironde, France. Private archives.

Initially made of wood, this fisherman shack, same as many more in the 1960s, had its walls covered with flat sheets coming from the nearby Everite plant of Bassens (near Bordeaux France). Roofs are made of corrugated sheets of the same origin. When I took this snapshot, I came to talk with people living there, some of them for fifty years, without visible damage to their health.
In the second half of the 20th century, a variety of corrugated sheets was created, apparently in Brazil, but with cousins in Europe. They were named *maxi-sheets* by some producers or *structurex* by others. Some were up to 8 metres long (8yds 2ft 3in) according to a hand free drawing and a photo extracted from the Brasilit Website Thickness reached was 8 mm (0,31in), that is some 30% more than conventional sheets, which gave them a high rigidity and permitted to use them as roofing in industrial buildings. In case of car parks, the need of very few posts to support the roof was a great advantage to move vehicles. Such sheets had a more limited success in France and were produced during only about ten years. In the contrary, they were very successful in many tropical countries.

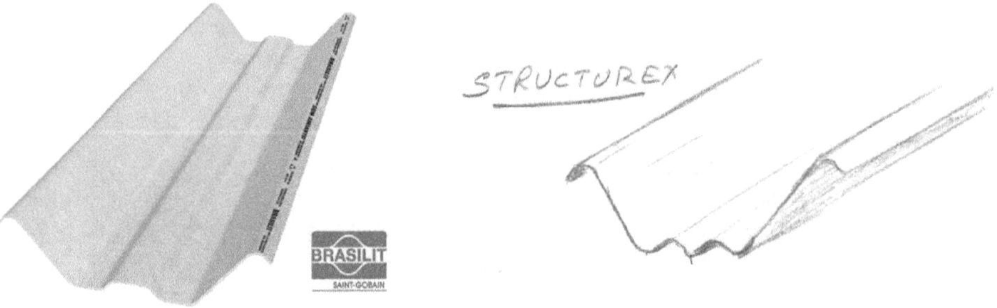

03-011. Maxi-sheet up to 8 metres long (8' 2" 3). Web Brasilit and private archives

The manufacturing of these large sheets required an adapted and impressive diameter of the making-roll on the Hatschek machine.
In Spain, they were made by Uralita in its factory of Getafe, Madrid, and called *Canalondas.* They had the particularities of one only wave, 7.50 metres in length (23ft), 0.90 metres in width (3ft), 8.5mm in thickness (0.315in) and 142 kilograms in weight (0.1565 short tons). They had been conceived to require just one post at each end. They were mostly used in rural areas to protect spaces dedicated to cattle, fodder or agricultural implements. They were also appreciated in very easy grade roofs.

As an anecdote it can be noted, that at the end of the 1990s when blue-asbestos was banned, Uralita built a shelter with such *Canalondas,* right behind the main workshop of its Alcazar de San Juan factory in order to store the Satanized fibre.

03-012 Everite Descartes. Indre et Loire, center of France. Note the length of the sheet and the required diameter of the making-roll. Private archives.

J.P. Guerber and his "Glossary of Asbestos Cement", mentioned several times in the previous pages, still offer a diagram of a strange and rare corrugating device which I think is worth to report here: Wet paste was dropped off on a train of belts which lead to the top points of the future waves. Belts were converging. It seems that the process did not meet a great success.

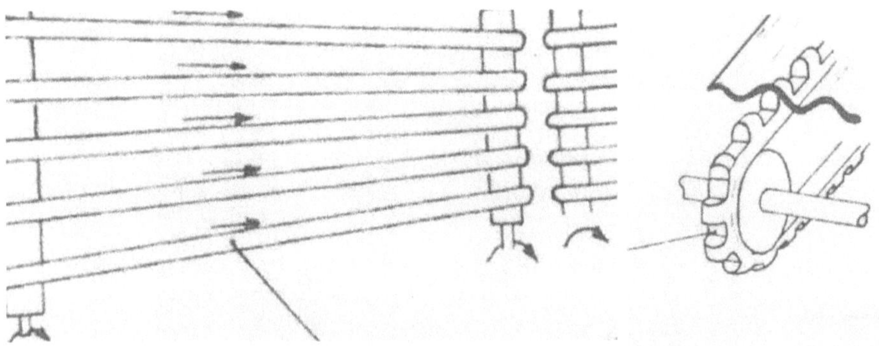

03-013 RCM Belt Corrugating Machine

Pipes:

"Make and sell pipes, was allowing men to live "

Roger Martin Patron de droit divin.

In the middle of the 20th century appeared a huge need of water supply and sewerage systems. The enormous investments required to set the various networks, justified search for cheaper products than the conventional pipes made of cast iron or concrete. As we could see, Italian Adolfo Mazza was the inventor of the first machine able to produce light pipes, which also meant easy to transport.

03-014. How easy to transport asbestos-cement pipes! "Uralita 80 años".

Those were easy to manipulate on work sites as shown by Uralita on his above advertising, and overall cheap to produce. If we observe the following diagram and if we remember the Hatschek machine making flat and corrugated sheets, we soon observe many common points between both.

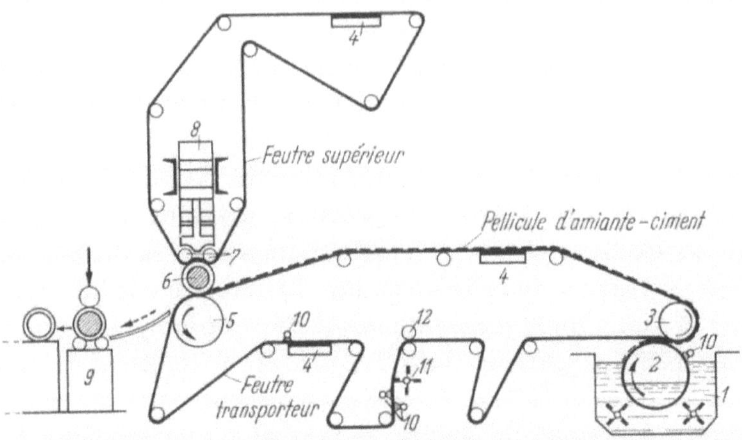

1Vat. 2 Turning sieve. 3 Couch-roll. 4 Suction boxes. 5 Motor roll.
6 Mandrel/making roll. 7 Pressure rolls. 8 Adaptive compressing devices.
9 Calender.10 Water sprayers.11Felt wiper. 12 Drying rolls

03-015. Diagram of a Mazza pipe machine. French speaking Switzerland Bulletin, December 1940.

Paste preparation belongs to the same technic, though percentage and quality of asbestos must be chosen and adapted to the type of pipes to manufacture: sewer-age, pressure, rain gutter downspouts, cable claddings, etc.

In the bottom part of the machine, one can note the presence of only one vat with its round-sieve, instead of the usual three or four vats of sheet machines. The paste is caught by the felt under the action of the couch roll, it loses part of its water when running over the suction box and then winds around the mandrel acting as making-roll.

The mandrel is a new device similar to the making-roll which we have seen at sheet machines. Its length is longer and its diameter is much smaller. The purpose is now to produce pipes and the pipes are much longer than wide. Technical problems are now beginning.

Separating the pipe, still soft and fragile, from the "making-roll-mandrel" without losing its roundness appeared as a delicate project. Adolfo Mazza's genius consisted in adding a compression device equipped with a hydraulic cylinder able to reduce pressure as the forming pipe was getting thicker. Adolfo Mazza registered two patents in 1922 and 1923 about his invention. In the early days, the length of pipes was limited to 3 metres (3yd 0.84ft x$7^{7/8}$ in) and the diameter to less than 20 centimetres ($7^{7}/_{8}$in).

The machine now described belonged to the second half of the 20th century. I never saw any belonging to the first half, nor could I meet anyone able to draw a reliable diagram. In any case, the many projects required could not be mechanised at the time, as they benefited later on, with help of numerous people working to improve the process.

Along years, lengths and diameters were highly increased in order to meet the market needs. In the 1970s, very big machines appeared to be capable of producing pipes up to 6 metres long and 2 metres in diameter (6,56yd x 2,19yd). This allowed a gigantic development of asbestos-cement pipes applied to multiple purposes.

But let us revert to the cylinder on which mono-layers winded one over the other until the required thickness was obtained. Let us remember that unlike the case of sheets there was no need to cut, but in the contrary roundness without any distortion had to be maintained. To achieve such a result the machine had to be stopped for a very short time. The pressing device had to be raised, the pipe and its mandrel expelled forward. As soon as the space was liberated a new mandrel was brought. Manually in the first times, then soon mechanically, and the pressure device had to come down. At the same time the machine restarted, the felt newly brought paste along to the new mandrel and followed its way to the cleaning system and to the round sieve where it pursued licking-up the thin mono-layer of paste in order to continue the process, en route for a new pipe, and so on. In the nearby pictures one can note the two felts of the Mazza machine. In the picture below, between the two felts, over the head of the worker who is watching, one has a good view of a pipe that just comes out of the machine and is still on its mandrel.

03-016 Note the impressive size and the pipes freshly separated from their mandrels

03-017 Over the head of the worker who is watching, one pipe still on its mandrel, is being taken out of the machine.

Then began a new delicate project: extraction of the mandrel out of the pipe without losing the roundness. In the 1950s/1960s, an air bubble was sent between the mandrel and the pipe through a flat nose pipette connected to a hose pipe generating compressed air. This small quantity of air allowed the mandrel swiftly slip out of the pipe. In certain cases, it was required to insert a wooden mandrel in place of the steel one, to avoid that the pipe loses it roundness, or even breaks during the following manufacturing steps. Due to their cost, steel mandrels were in limited quantities and reserved for the forming part, a reason why they were replaced by cheaper wooden mandrels, which could easily remain as support for hours. The steel mandrels could then return to where they were needed and start a new circuit.
With modern machines, built after the 1970s, the air bubble solution was replaced by what was called an "electrolytic calender" which consisted in sending 400/500 volts direct current through the thickness of the pipe, by means of a kind of cylindrical graphite brush. This system would lightly unstick the pipe and the mandrel, permitting to extract the mandrel with help of an electronic piston.
Once separated from its original mandrel, possibly replaced by a wooden one, the pipe was transported, rolling on a whole set of rolls, to the following manufacturing steps.
And the rhythm was the same, days and nights, except when some incident occurred and disturbed the duly programmed "ballet of mandrels and pipes". In that case, one immediately had to stop the arrival of paste, to urgently clean the probable excesses of paste with high pressure water jets before they start hardening and send them back into the circuit. Then people in charge had to find out the origin of the trouble and solve it before restarting production. During that time, fresh pipes just freed from their mandrels, followed their route towards the stove and stood there a few hours. Then they had to cure from 21 to 28 days in the open air, while being steadily sprayed with water by automatic systems. Some manufacturers, such as Uralita in his plant of *Alcazar de San Juan* would proceed differently. They stored their pipes for curing time in very big water basins which looked somewhat like Olympic swimming-pools.
Another pipe manufacturer in a remote country, when making pressure pipes, made them stay some 8 hours in a 170°C (248°F) air-steamed oven, which noticeably improved their quality.
Once the pipe curing was over, the pipes still remained in the workshop, so that their ends and edges were perfectly clean to enable finishing.

03As with other types of pipes, the inside was sometimes covered with a protecting coat of varnish, for instance bitumen, in order to avoid corrosion due to sulphuric acid resulting from certain waste waters.
Another important finish consisted in preparing the coupling of pipes in order to foresee their installation on work areas.

Pipe coupling:

Fig. 72. Adoucissement à la râpe de la tranche d'un tuyau.

03- 018.Worker machining the edge of a pipe. Kurt Hünerberg

Setting pipes anywhere requires to be able to connect them together and do one's quick as possible. Consequently, the ends of the pipes had to be polished and machined. The type of junction had to be chosen according to the type of pipes and their final use. When pipes were intended for water supply, junction consisted in: a cylindrical sleeve "sliced-out" of an asbestos-cement pipe with an inner diameter equal to the outside diameter of the pipes to connect- The length of "slice" was from 10 to 15cm (4in to 6in) and the inside was machined in view of receiving the adapted leak-tight seal -waterproof rubber seals were usually purchased from subcontractors.

When pipes were intended for building construction, rain gutters, cable or air ducts, rubbish chutes, they were most of the times fitted in one end with joining sleeves stuck in with epoxy resin by the pipe manufacturer

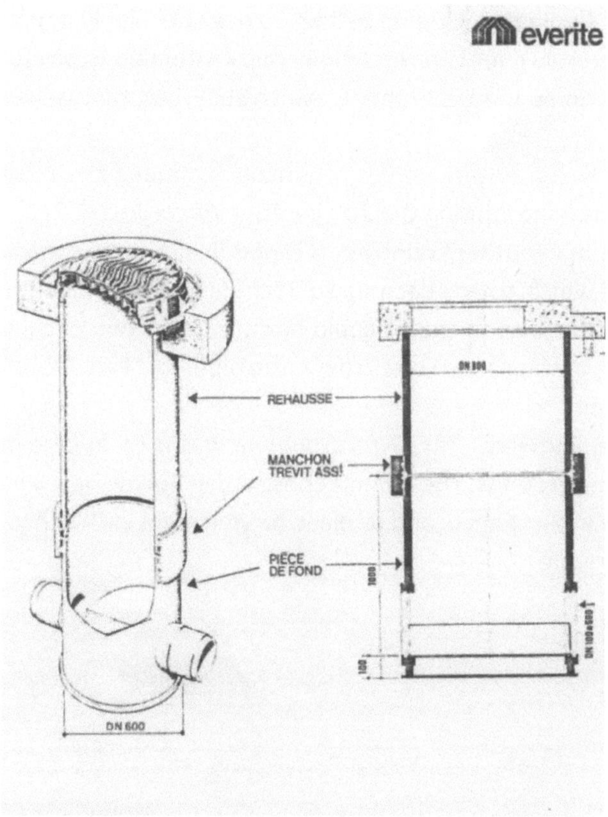

Installation at work areas was much easier. Some works required elbows, bending and manholes, which asbestos-cement manufacturers were also able to produce. Some companies, such as the Spaniard Uralita, were using Gibault tight-joints which had to be settled on the work area when dealing with short diameter pipes.

Pipes bound to be buried, water supply, sewers or rain waters, were most times delivered "naked", that is with their joining system apart but ready to be installed on the work-site where they were press-fitted.

03-019. Everite sewerage accessories. Private archives

When a miner's pick was not sufficient, one asked for help to the nearest digger. In case of a straight pipe line, heavy pressure was easily accepted. In case of a bended pipe line, without sharp angles that would have required other accessories, equally made of asbestos-cement, it was necessary to mechanically latch junctions, otherwise pressure could have unfastened the pipes, which absolutely had to be avoided. In France, on its side, Everite developed its own coupling and seals, as shown in the next drawings.

Burying pipes in trenches had to be performed according to a duly definite procedure. It drew asbestos-cement makers to publish accurate user's manuals and even to organise training sessions so that their instructions were respected accordingly.

Such sessions often turned into seminars including certain "blow-out" highly appreciated by guests. Some manufacturers even offered scholarships, intended to pay for the training of site supervisors, in specialised schools which they had themselves founded and paid for.

In the following page, you can find a few examples of junctions, among the most commonly used ones for pressure pipes during the 20th century. *(diagram 03-022)*

Frequently used in Spain and in many other countries, a *Guibault* sleeve was made of an asbestos-cement cylinder which could reach up to 35cm ($13^{25}/_{32}$in) in length. The inside was machined in the factory, so that it could be fitted with two circular leak-tight joints and required separating accessories to put the pipes in a line, including in case of bent up to 5°.

A variety of the *Guibault* sleeve, known as RKT *(T for traction),* was used in case of sea outfalls[55] or on highly slopping grounds. The technic consisted in hollowing a kind of channel, both in the pipe and in the sleeve, and to place a nylon stem in the channel, so that pipes could not uncouple.

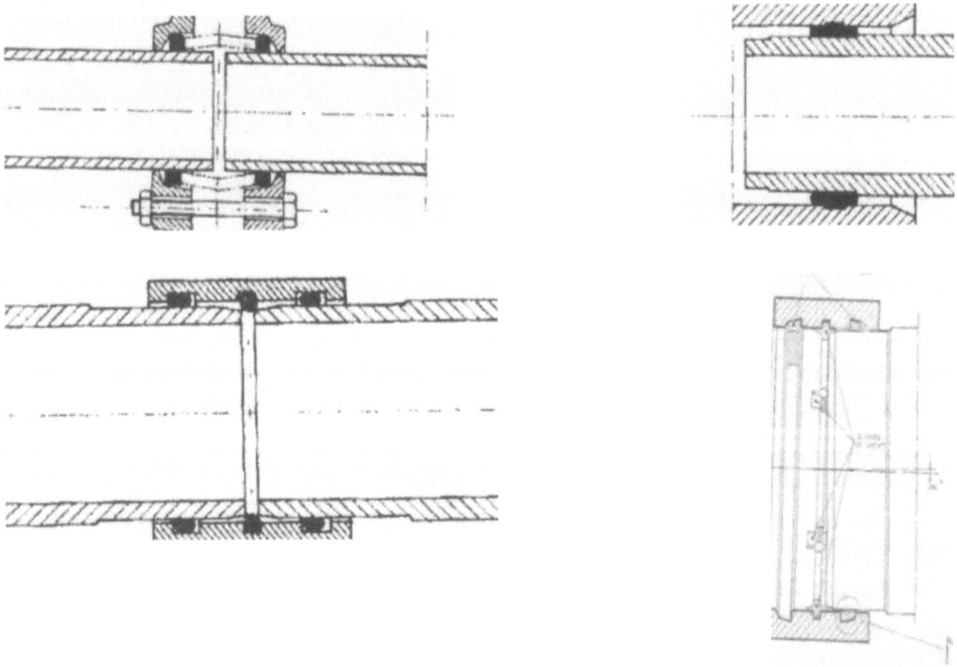

[55] Sea-outfall: large submarine pipe-line used to carry waste waters as far as possible from the shore.

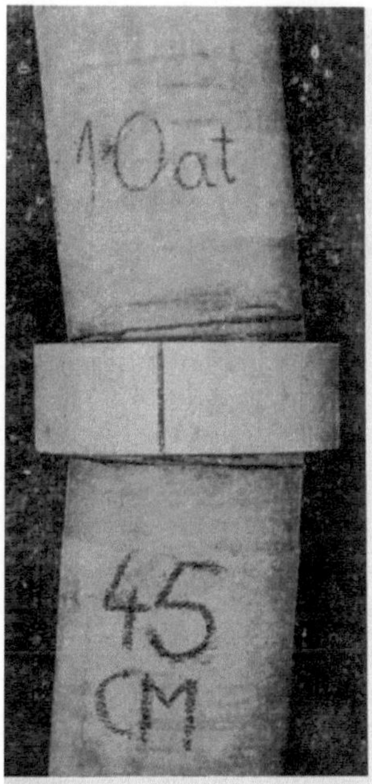

'ig. 31. Joint REKA reliant deux uyaux défléchis. Les traits noirs ndiquent la pénétration dans le nanchon quand les tuyaux sont alignés.

03-022 03-023 03-024 03-025 Pictures extracted from "Asbestos -cement pipes" by *Kurt Hünerberg*

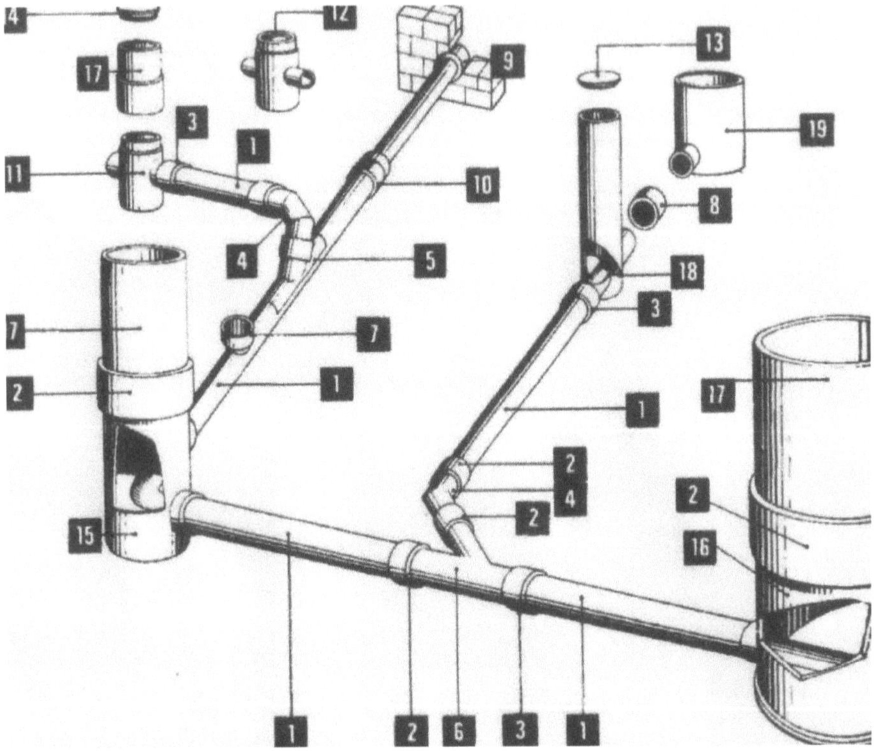

1 Processed or un processed pipes end
2 F or T connecting sleeve
3 Evermetic connecting sleeve
4 Elbow
5 Connection with slab
6 Simple connection
7 Access connection
8 Closing sleeve
9 Pull-up sleeve
10 Cone
11 Transfer connection
12 Waterproof connection
13 Ordinary cover
14 Male cover
15 Vision panel
16 Manhole
17 Manhole access
18 Direct connection box
19 Connection vision panel

03-026 Extracted from a 1984 Pont-à-Mousson catalogue showing a complete system of accessories such as manholes and all kinds of connecting devices, all made of asbestos-cement. Pierre Picavet archives.

03-027 Test work. Pierre Picavet archives.

Before leaving pipes, let us have a look to some more machine inventors, who tried, and sometimes succeeded, to compete with Adolfo Mazza.

For instance, Magnani developed a one felt machine, which seems to have supplanted Mazza in many companies and countries, after having been more or less improved by the various users.

In the first half of the 20th century, Magnani created the T Machine according to the following diagram[56] It was a very simple machine which was used in France until the 60s. It consisted in a monolith mandrel, with vacuum interlocking covered with a cloth. Paste was intentionally thicker than the one used on a Mazza machine, it was laid between the mandrel and an equalizer roll with a complementary profile, running opposite direction.

03-030 Magnani T Pipe Machine.

(Former process)

Called Magnani T (T for Toile in French, meaning cloth).

Supplied with pulp.

Pipes without single layer.

Pipe length: 3 to 4 metres.

Pressure or sewerage pipes.

Monolith interlacing.

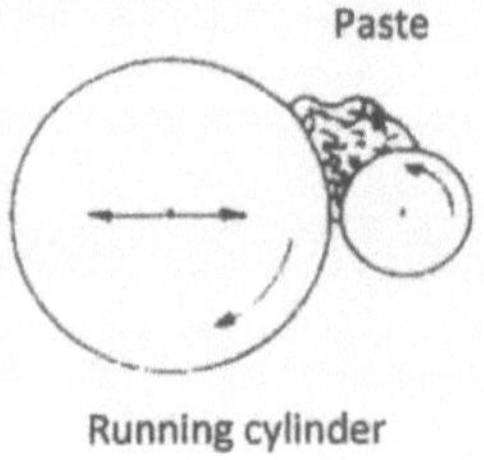

Not only was the size of pipes limited, but quality was rather poor due to the three directional fibres orientation, which divided by two the endurance compared with Mazza machines and their bi-directional oriented fibres.

[56] Extracted from "Asbestos-Cement dictionary » by J.P. Guerber, year 1963. J.P. Guerber was Research Manager at the French Everite from 1962 to 1969. We owe him lots of technical information about ancient machines and technical vocabulary appearing in the present book.

Never short of new ideas, Magnani did not stay on his primitive T Machine. He invented a second generation which, coupled with numerous improvements offered by its users and very convenient for small diameter pipes, knew a brilliant life.

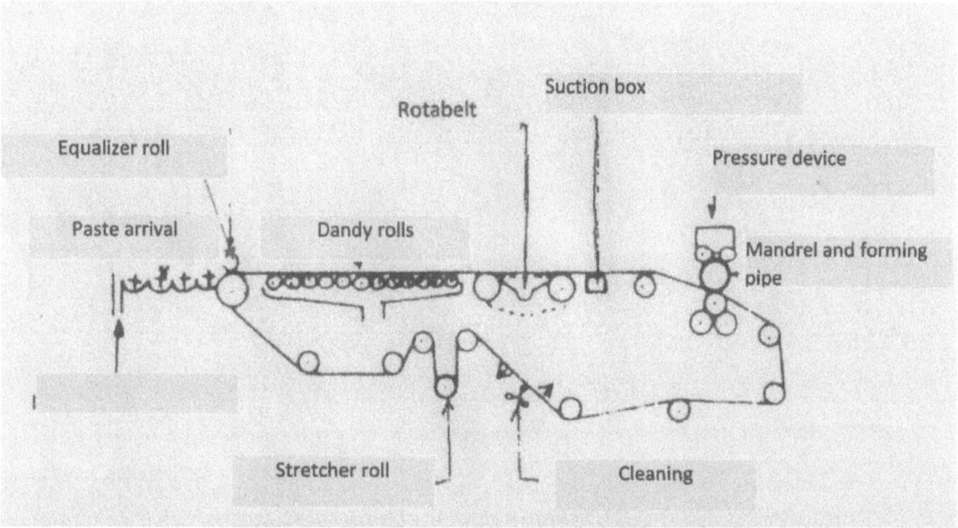

03-031 Magnani T new Process. JP Guerber.

Dalmine, another Italian, invented a pipe machine using a very special technology. The mandrel did not just turn on itself as it did in other machines, but it also had a parallel motion on its own axle. The full of imagery vocabular of users, often called this type of machine as "puttees machines"[57]. It seems that the success of this technic was rather limited. According to the inventor, the interest of this machine was to be able to make pipes in different length. In fact, the motion of the mandrel allowed a helicoidal deposition of the single layers, hence the likeness with the puttees...Pipes made with this machine were very flexible and presented quite a good resistance to tensile strength. Some of these machines were fitted with 3 to 6 felts without reaching a sufficient production for a good profitability. The French Everite

[57] Puttee: A long strip of cloth wound spirally round the leg from ankle to knee for protection and support. Was part of the uniform of many soldiers during WWI and WWII.

used one of them in his factory of Dammarie-les-Lys, near Melun, Seine et Marne, France, during a few years.

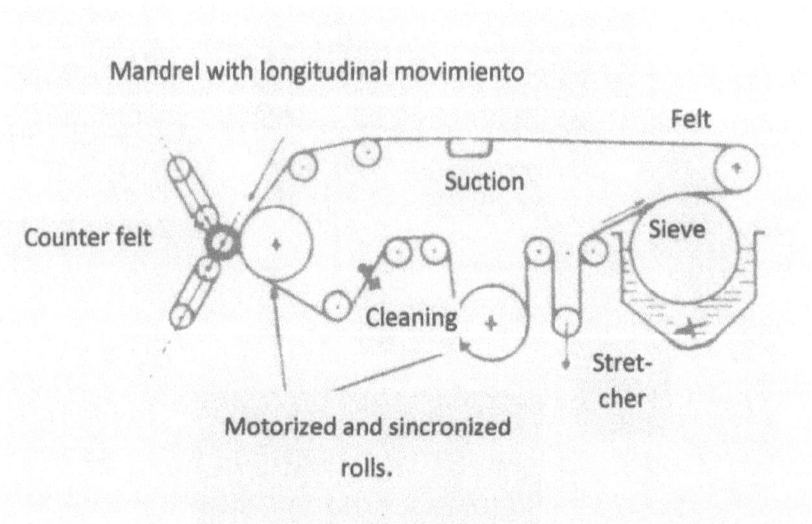

03-032 Dalmine pipe machine. JP Guerber

03-033. Puttees on the legs of a soldier,
most likely in the beginning of WWII. Internet picture.

Hamanit, another inventor, conceived a machine corresponding to the following diagram. It led to a Swiss patent, which met only few successes among asbestos-pipes manufacturers. As you can see in the drawing, the paste was directly dropped loose on the felt, was then wound on the mandrel. A guide roller on the felt, acted as a tipping device, which according to its position, permitted to enfold or remove the mandrel[58]. This machine had the particularity of being able to produce concrete pipes, and it seems that it knows a certain revival in the beginning of the 21st century, in the manufacturing of pipes without asbestos. Unfortunately, I had no opportunity to check this last point.

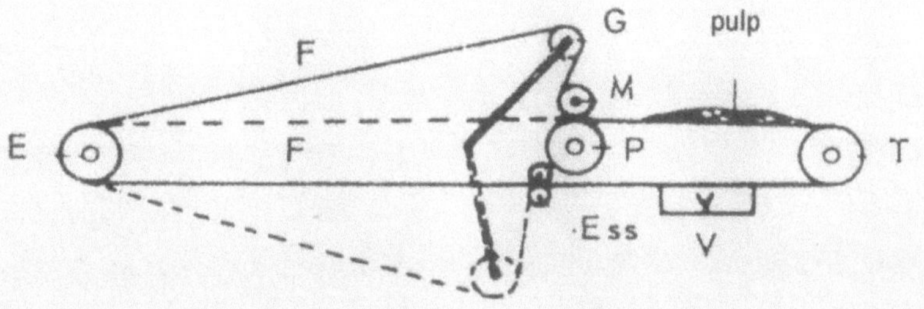

P: pressure - E: more pressure- V: suction box.

03-034 Hamanit Pipe Machine

Finally, being given that engineer's cleverness is unlimited, each company once the manufacturer's warrantee was over, enjoyed to improve the capacity of the machine it had purchased and considered to have created the best tool in the world. Some asbestos-cement manufacturers went so far as to conceive and produce their own machines in order to set them in their branches and even to sell them on other continents. In turn, the new purchasers hurried-up improving their new tools.

[58] The various diagrams above about these types of machines are all extracted from the dictionary of J.P. Guerber already mentioned.

03-035 Everite Andancette Factory, Drome, France. 3 metres flow-on pipe machine AT3, built 1965. In the foreground note the stirrers of the distributor, typical of a flow-on machine. Unfortunately, the sieve situated under the felt is invisible. Private archives.

In this case one can see in the Everite factory of Andancette, Drome, south of Lyon, France, a 3 metres wide pipe machine, built in 1965, equipped with a double paste supply. By a sieve under the felt and via a distributor or feeder on top. From the users of this machine, the top feeder was practically never operated. The truth is that it had been added in case of need, as this machine was the first pipe machine in the company with paste delivery through a sieve. In the second half of the 20th century appeared a quite specific use of asbestos-cement pipes.

03-036. Irrigation canal made with half big asbestos-cement pipes. Private archives

This was particularly developed in Spain and some other countries, where nature requires important irrigation, justifying big networks of water supply on large distances. The system consisted in cutting big diameter pipes, 800/1000 mm, (31 ½ /39 3/8in) according to two diametrically opposite straight lines. Settled end to end, sealed one another and most times supported by concrete stands. They would form large canals running through agricultural plains and bringing water where it was needed. Combined with plastic greenhouses, which appeared more or less at the same time, they gave the opportunity to Spanish horticulture to develop the country thanks to its exports of fruits and vegetable through most of Europe, sometimes to the great dismay of neighbouring countries. The main enemy of this type of network was the sun which generated high evaporation. For this reason, they soon gave up the process.

SEWERAGE

Fibre-cement manholes for sewerage works

The lack or shortfall of waterproofness in some nets is a serious worry for public authorities, due to the economic and ecological consequences, especially:

-useless additional work for treatment plants for parasitic waters, generating important excess in energy expenses, which put heavy strain on the often-precarious budgets of local authorities.

-environmental contamination, particularly in underground waters, which is the main source of supply of drinkable water.

-The lack of waterproofness of subterranean nets, results for a great part of the lack of waterproofness of manholes required for control and maintenance.

EVERITE Co, offers a number of new manholes, from 600 mm (23 5/8in) to 800 mm (31 ½in) in its range of EVERMATIC.

Such manholes, made of standard fibre-cement pipes and junctions entirely prefabricated grant a perfect waterproofness to works and to their junctions with sewerage nets, under inner and outer pressure up to 1 bar.

DN 600 cleaning manholes can easily by used instead of conventional manholes, up to 1,50 metres in depth. In case of depth over 1,50 metres (1,64yd), EVERITUBE suggests to use DN 800 manholes.

The use of standard junction pieces to connect them to nets, as well as their relative lightness allow, apart of easy settings, an almost immediate backfill of trenches (specially interesting in case of under road works).

Moulding

Fresh sheets, just cut off the forming cylinder, either on Hatschek machine or on flow-on machine, are very malleable as can be seen with corrugated sheets. Consequently, the idea of moulding this paste to create all kinds of things was evident from the beginning.

03-037. Two workers are rolling fresh paste which will soon be moulded.

The process consisted in removing a given length of flat paste just fallen from the making roll and to let it wind around a cylindrical support held by two workers who carried it to a table where moulding could be performed under acceptable conditions.

In the first times, moulding activity was nothing but an accessory. Moulded parts were needed to sell slates and corrugated sheets. Making corrugated sheets and selling them to roofers was a good thing, but offering the corresponding accessories such as ridge and edge tiles or chimney tops, sounded much better.
Large series mouldings began in Great Britain and in the Netherland's where Eternit met its first big successes in the field. Perhaps thanks to the fact that unlike metal sheets, asbestos-cement sheets were not damaged by rain corrosive effect. Such success led to new accessory mouldings.

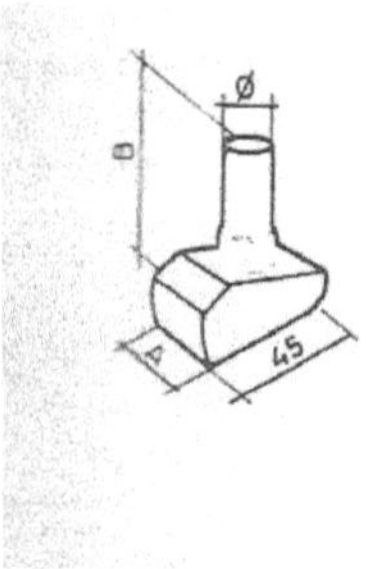

03-038 Chimney top

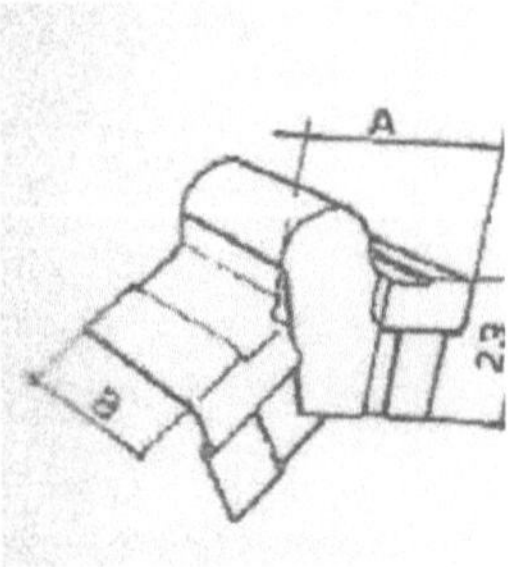

Ridge Tile

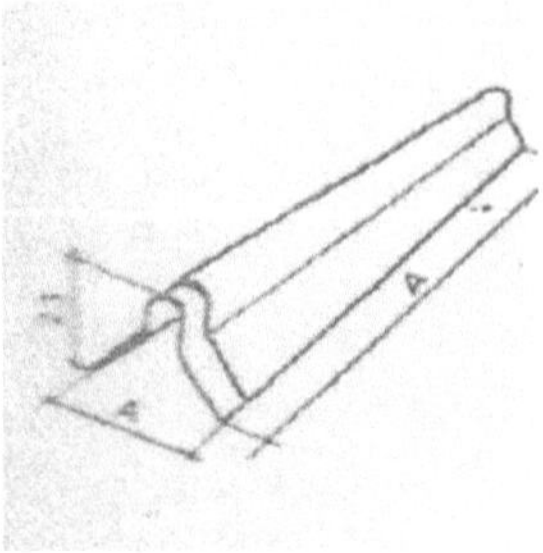

Edge tile

The first moulded pieces not tied to sales of sheets, were seed-trays produced in Switzerland as early as 1914. Soon after appeared asbestos-cement flower stands, sanitary pipes as well as air and cable conduits.
Provided that water percentage is carefully chosen, fresh asbestos cement paste is perfectly appropriated to all kind of slicing and shapes.
Using patterns, even sometimes cookie-cutters in shape of the object to manufacture, one would cut a surface of flat paste corresponding to the surface area of the piece to be produced. Some objects needing a perfectly smooth inner surface required an inner mould. Others, with twisted shapes required outer moulds, and would it only be to facilitate demoulding.
Most times, inner moulds were made from 3 parts. On picture 03-038 the mould was conical, making demoulding quite easy. The outer mould was made in 2 parts which, once cement hardened, could be peeled away like bark or peel. In the case of some particularly complicated pieces, it was occasionally needed to use up to 4 different parts.

Moulds were made of wood, of asbestos-cement or metal. Some pieces required several welds. The two ends of the piece of paste were manually made thinner in order to avoid any extra thickness on their junction. They were then welded by a vigorous hammering. In case of heavy pieces, such as the square duct of pictures, the beating surface of the pneumatic hammer was very rough in order to really knead paste. Fibres were then so strongly interlocked that the area of a well-made weld was equal to that of any other part.

03-039 Inner moulding

The endurance of moulded items varied according to their profile. The simpler the piece, the more enduring it is. In return, the more complex, the more breakable is it.

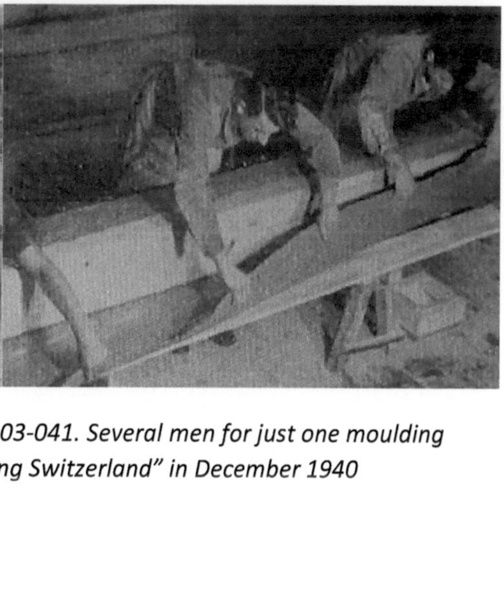

03-040. Welding with pneumatic hammer. 03-041. Several men for just one moulding Technical Bulletin of French Speaking Switzerland" in December 1940

Sometimes, success required to add some water during the process. The entire work required an adapted organisation as the whole had to be completed in one or two hours after the sheet fell from the machine. That meant before the start of cement setting.

Small pieces were most times entrusted by women. However, sometimes the biggest pieces required the strength of four men, so that the work could be completed without destroying the structure of the asbestos-cement.

One can say that this industry shares certain points with the foundry. Indeed, they both need previous and serious studies, as well as great experience, in order to determine the best adapted shapes for the pieces to manufacture[59].

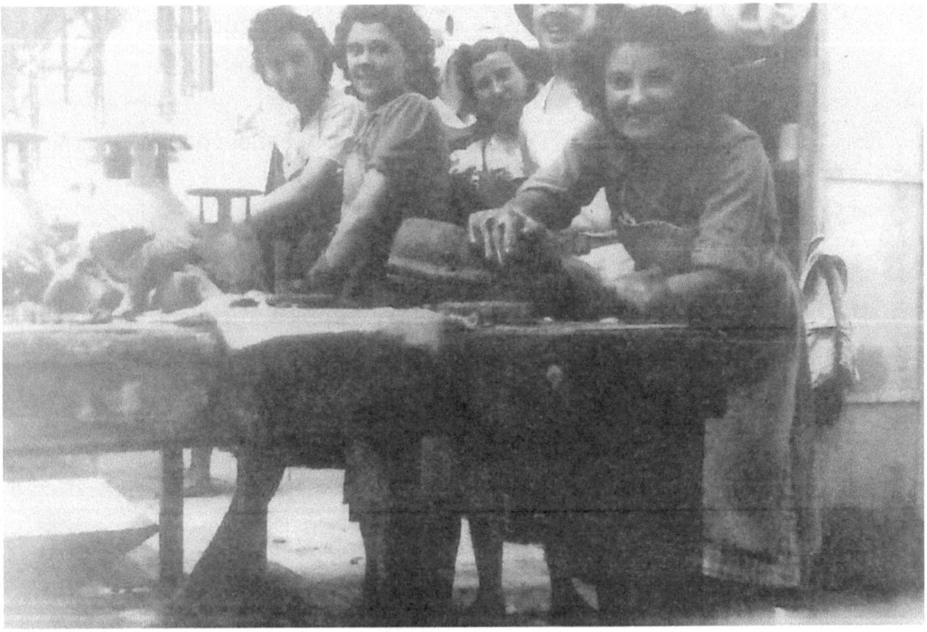

03-042. Ladies working on mouldings. Everite Bassens Factory, near Bordeaux, France Years 1950.

[59] Explanations and pictures of moulding, are mostly based on an article of H. FREY, Engineer at Eternit Niederurnen, Switzerland, published in the "Technical Bulletin of French Speaking Switzerland" in December 1940. If the described technics belonged to Eternit, one can bet that competitors were quite similar.

III Manufacturing process and products

Let us have a sight through the eye of Ray, operative of this activity who tells us his memories of the 1940s, when he was operating in British Uralite at Higham, Kent, UK:

"We were manufacturing window boxes, which entailed multiple trips to the hopper, where sheets of wet asbestos were laid out on giant tables, about 6 ft x 8ft long. Using a giant rolling pin between two men, we would roll a sheet onto the rolling pin and carry it back to our work station, there it would be unrolled on the table and various sizes would be cut from the wet sheet. These would be placed in wooden moulds and beaten into shape with a metal spatula.
These moulds would be placed in a rack overnight.
The following day these moulds would be back on the bench where the semi hard asbestos would be trimmed with a knife or hack saw blade, the mould placed back in the rack.
The following day these moulds that had been trimmed would be dismantled to remove the window box, these window boxes would be stacked on the floor around the work area, where they would be left to cure or harden.
The next day the cured window boxes could then be filed, sand papered and all edges cleaned, they would then be dusted down with a powder where they would be taken to the stores.
With all the filing and sand papering being carried out this inevitably created asbestos dust, which settled on the benches, on the floor, on your clothes and all other sections in the open factory area. The same dusting, cleaning, sawing and sandpapering created dust everywhere and on everything as no attempt was made by the management to clean the area. In the event that the window boxes or other asbestos items were damaged and acceptance was refused by the stores, they would be broken and the pieces thrown in the bin.
When there was a strike you had to tear open the ready filled bags of asbestos for the hopper in an enclosed area in which asbestos dust and fibres were over all the machinery, up the walls and about half inch all over the floor. The air was thick with dust. At no time did British Uralite offer any breathing protection nor any warnings".[60]
In the 1960s, manufacturers were searching for solutions to reduce costs of manpower considered as expensive in mouldings. Two Italian companies and one Spanish

---

[60] A history of Higham Roots, Mickle Print (Canterbury) Ltd 2011.

company finalised a mechanical process. It consisted more or less in high-pressure injections of a quite concentrated asbestos-cement paste into a mould, the outer wall of which was a filtrating unit *(canvas cloth, sieve and perforated plate)* connected to a suction device. A compressed air setting and an inflatable bladder conceived to fit the mould, permitted to put the paste in the desired shape. The only remaining operations consisted in demoulding, then curing and polishing.

The most famous moulding presses were the Italian Sambuco and Essam. In France, Le Fibrociment de Poissy and some Eternit factories used several of them, only for small mouldings with easy shapes, avoiding complicated de-moulding. It seems that results were lukewarm. Some countries, such as Spain and Brasil made a large use of the process. In our beginning 21st century, a Canadian inventor filed a new patent. Up till now I could not get any information as to the result of his system.

Before leaving moulded products, let us have a look at something quite unknown from Europeans, but very frequent in countries with less luck as to water availability and supply of the so vital commodity.

I discovered those *"tinacos"* in Mexico. Yes, that is how they name these big water tanks which you can observe on the roofs of lots of houses almost everywhere in Latin America. They have a cement grey colour. Why? Simply because up to the end of the 20h century, they were made of asbestos-cement[61]. In many places, companies in charge of water supply can ensure deliveries only a few hours per day and per area. People seize the right moment to fill up their tanks. One of my friends, an engineer who worked the main part of his life in a big Mexican company, at a branch of the French Saint-Gobain Pont à Mousson, explains for us the manufacturing process.

Tinaco (water tank)

[61] By now, asbestos-cement has been replaced by plastics and most "tinacos" are black.

Manual process

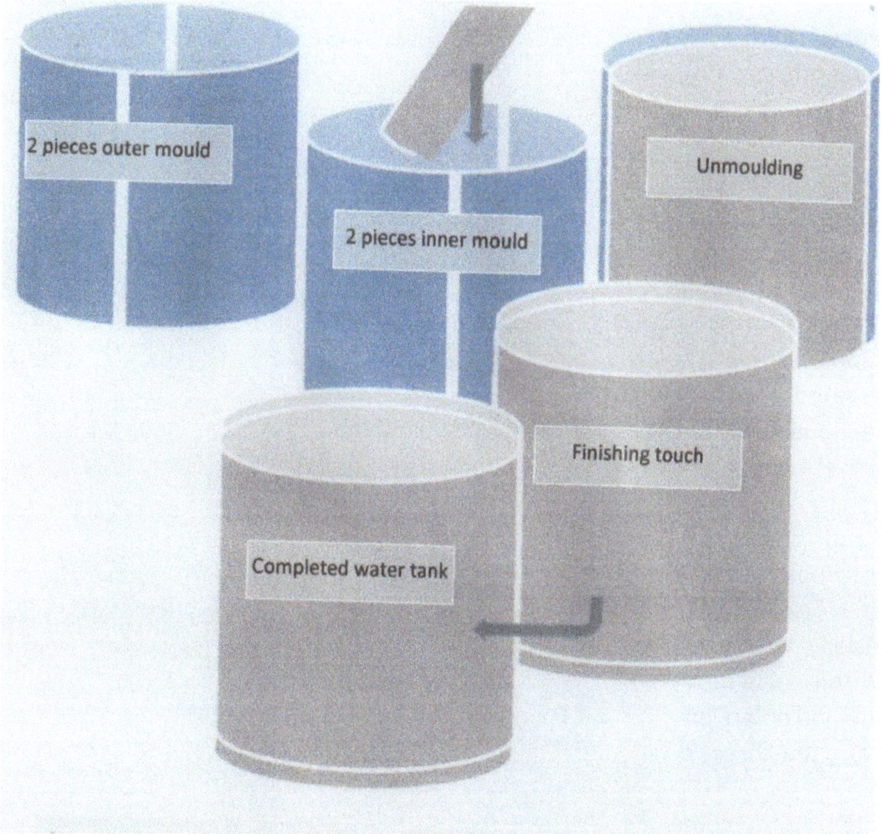

Shapes, most times cylindrical, but also square, or mix or even different.
-Machine: no
-Moulds: in two parts, made of asbestos-cement, or steel, or aluminium.
-Raw material: fresh asbestos-cement sheets hand collected from the machine and
-arranged inside each of the two parts of the mould.
-Careful hand compaction in each *03-044 Tinacos: manual process by Fernando de Aragon O*
half of the mould.
-Gathering and welding of both parts with extra-compacted paste.
-Recovery of the mould after two hours, then hand smoothing to eliminate burrs.
-Air curing for two weeks.

Mechanical process

-Shapes: Same as manual process.
-Machine: yes
-Raw material: semi-fluid high-consistency asbestos-cement paste.
-Supply: by pump.
-Distribution: homogeneous through a perforated vertical cone, doubled with a filtering cloth, animated by continuous up and down motion, coupled with a suction device.
-Specially designed steel rolls, also moving up and down, improving the compaction.
-Other rolls compacting the bottom and welding of the latter.
-Finishing: smoothing with semi-manual tools.

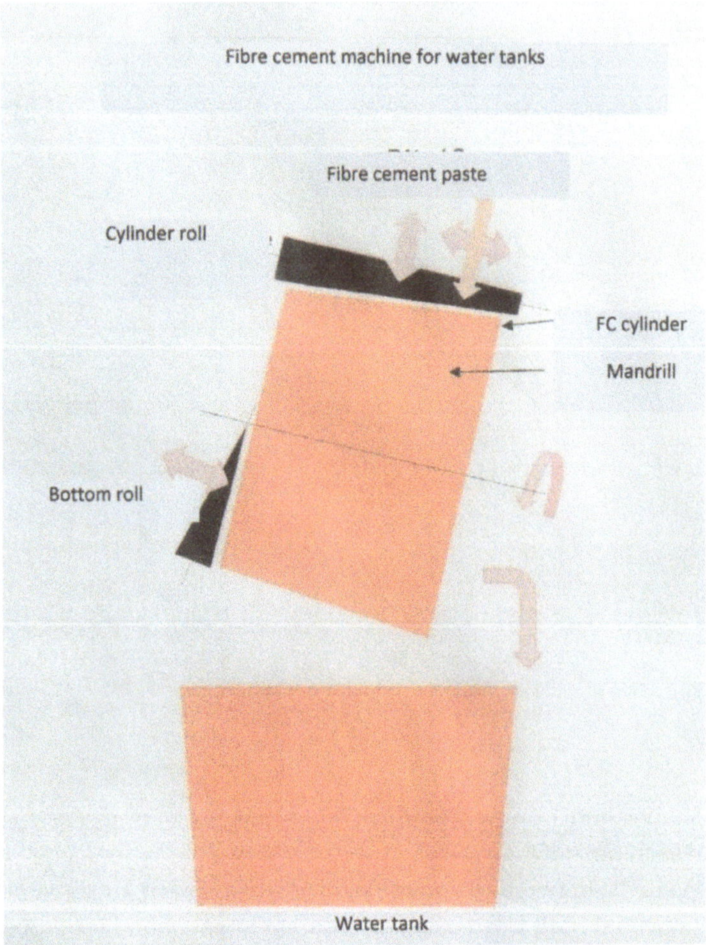

03-045 Tinacos, mechanical process by Fernando de Aragon

To come to an end to theses technical points regarding asbestos-cement, we can have a new look at the global diagram -already present on the first page of the chapter- showing the complete manufacturing circuit of the main products, from arrival of raw materials to shipments. One will observe that the left part, regarding paste preparation, is common to all of them. The right part sums up the particularities of each group.

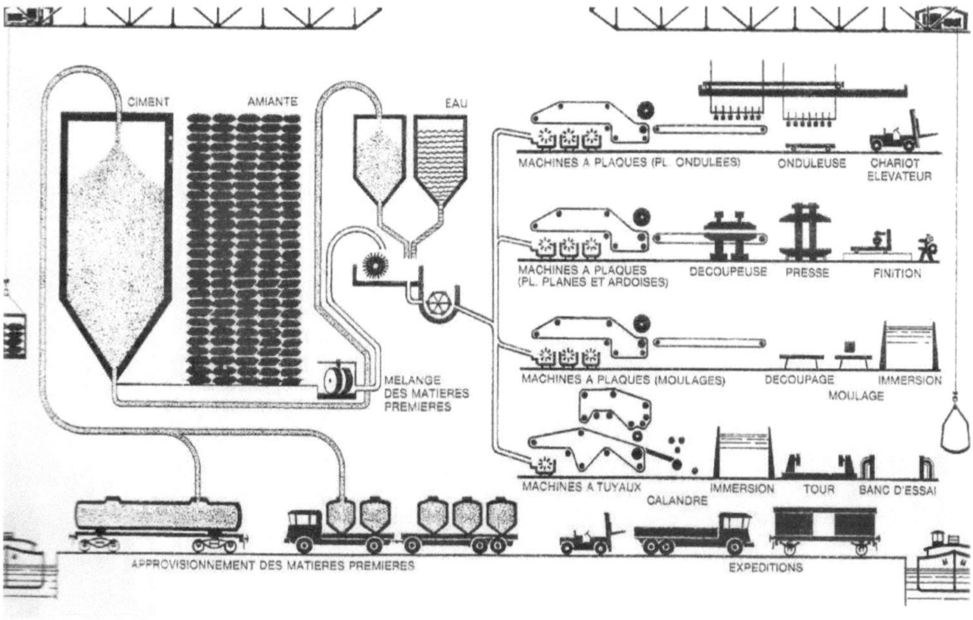

From top to bottom

- *flat sheets,*
- *corrugated sheets,*
- *pipes,*
- *mouldings.*

To any reader wishing to know more about the above, I suggest a visit to InspectAPedia's website:

Asbestos Cement Production Methods®

Asbestos cement roofing, siding & millboard composition, history, production, uses, identification & safety .

IV Worldwide outburst

Though inner and outer decorative boards existed before the Second World War, their intensive production began only after 1945[62].

1945, WWII was coming to the end. Most of Europe was destroyed and some countries in other continents were hardly better. The process of rebuilding was urgent: means of communication, public buildings, water supply networks and dwellings.
Factories had also been greatly damaged by bombings. Such was the case of the factory of Bassens, near Bordeaux, southwest of France, in spite of its intentionally localisation far from arm fighting areas at the time of WW I. It was among the first ones to be rebuilt and still named *"Sociétés Réunies Everite Situbé"*. Already in the first months of 1946, production restarted slowly. The country's general situation was an obstacle to acquire machines and raw materials. In spite of such situation, thanks to the activity of people in charge, restart was faster than expected. If this factory counted 500 workers in 1939, in the years that followed the end of the war it reached 1160, and modern machines improved quality and production rates.
In Northern France, they also hurried up to reset their machines in operation. Such was the case in Thiant, where the French Eternit was established more than 20 years earlier. Pre-war production was soon doubled, yet profits did not grow at the same level due to the general inflation rate and to heavy investments required. Working conditions were still rather difficult. Two new 5 metres pipe machines were built and structures were set up in order to improve storage and maintenance. Yet, working conditions remained painful: *"Workers are in their wood clogs and when the sheet*

[62] Unknown origin.

reaches the right thickness, it is hand cut with a big knife" reports a memo written by Jacques Decret on behalf of a local heritage conservation organised by a regional museum.

04-001/04-002 Bassens Factory-near Bordeaux France- at the end of WWII

Development was not restricted to mainland France. New factories were planned, first in Algeria and Morocco, but soon in faraway countries. In Kouba, south of the town of Algiers, production began as soon as 1950 with two sheet machines and two pipe machines. Eternit already conceived asbestos-cement pre-made houses called "Proplac", bound to be built in Sahara Desert. Not to be outdone, PAM[63] settled the project of a factory next to Oran, northwest of Algeria, naturally named Oranit. Due to the persisting needs in raw material and to the increasing demand of rebuilding market, as soon as 1947, French and Italian producers joined to develop the asbestos mine of Canary, northeast coast of Corsica. This mine came to a consequent production. But finally, the quality of its asbestos was considered as poor, which lead to its closing as soon as 1965. The mine caused lots of victims and an awful site which neighbours would like to make disappear, even if hereto, one perceives some nostalgy of the time when it offered work to so many people

Loup Espaguillere, le Monde.fr 2015 09 02

04-002 image Thierry Santini, le Monde 2nd September 2015

"The Asbestos manufacture of Canari, Cap-Corse, North of Corsica Island, was definitely closed on June 12 1965. Fifty years later, this industrial wasteland and its quarry go on, year after year, to waste millions of Euros public money, which are spent with one only purpose: reduce environmental hazards". "On hill slope, besides the picturesque coastal road, between sea and mountain, the old buildings of the factory nowadays seem quite difficult to clear of asbestos, and even difficult to dynamite. Armand Guerra, mayor and formerly clerk in the company from 1959 to

[63] PAM: remember is the familiar name of the famous French firm Pont à Mousson, specialized in pipes pre-manufacturing.

1964, re-elected as mayor in 2014, warns: "if we destroy the place, asbestos fibres will fly away everywhere. Leave the buildings as they are. This is the only way to protect us from the poisonous fibre".

In metropolitan France, at the same period, Eternit was improving its painting technics on sheets and its ability to extract mandrels from fresh pipes.

In 1946, Canacit was created in Caracas, Venezuela.

Between 1947 and 1949, Eternit Switzerland, in its Niederurnen factory, proceeded to tests and advanced research on pipes foreseen to be possibly used in hydroelectricity. There were fears of a lack of pipes for this activity.

In 1949, the French Eternit was buying shares in Dimatit near Casablanca in Morocco. Everitube-Situbé, with its factory of Bassens, near Bordeaux, then totalising some 1300 workers, was selling great part of its production in what France still called "the Empire", mostly: Indochina, North and Equatorial Africa. Such sales were classified as "exports".

On May 10th,1950 PAM created a new company called Perdurit in Cuba. Some Magnani pipe machines were built there in 1953. The company was nationalised by Fidel Castro when he took power in 1959. This did not prevent PAM to sell new machines and know-how a few years later. We will revert to this point in chapter VIII.

In this mid-20th century, Le Corbusier, famous French architect, was using big quantities of asbestos-cement roofing and cladding when fulfilling its building projects. Again in 1950, a member of PAM, *(whose name could not be found with certainty)*, published a memo about asbestos-cement in France and in the world. Apart of many technical data, one can read:

"France holds a favoured rank in natural building material widely available on its territory. One can think that the invention of an artificial material should not create a great need, and that consequently, asbestos-cement should not meet such a big success as in other countries less fortunate in this field.

Yet, jobs in the industry of asbestos-cement are widely developing. By now, the French output is likely between 80 000 and 100 000 metric tons (88 200 to 110 200 short tons).

IV Worldwide outburst

	In square yards	*in short tons*
National consumption:	*11 522 526*	*96 450*
Exports:	*730 591*	*6 063*
Total:	*12 253 117*	*102 603*

Asbestos-cement is now produced in the following countries:

Belgium	*Sweden*
United Kingdom	*United-States of America*
Austria	*Canada*
Checoslovaquia	*Brasil*
France	*México*
Spain	*USSR*
Italy	*Japan*
Switzerland	*India*
Germany	*Australia*

On 8th December 1950, Joseph Cuvelier, who founded the French Eternit, died. The workers of the company expressed their gratitude for his industrial and social mind. His son, Guillaume Cuvelier succeeded. He continued his father's work and in the following months new pipe machines were set in. During the years 1952/53, Roger Martin then Vice Director of PAM, was trying to counter the French Eternit, whose dominant position tended to increase, as the Swiss Eternit was rising its power on the family members in other European countries. This is how PAM was foreseeing the development of asbestos-cement, and other industries using asbestos, but un-satisfied with the "small mines "of Canari, Corsica and Bom Jesus, Brazil, PAM was searching for new sources of supply. A few years later their research were fulfilled with the discovery of a mine in Cana Brava, Brazil by a young engineer, who he had contracted in this very purpose. We will revert in details on this event, as well as on the birth of "Productos Mexalit" in the same years, also under guidance of PAM, when writing about the American continent.

In 1954, on the initiative of Philippe Cavalier, belonging to the famous family who founded "*The Fonderies de Pont-à-Mousson*" (Pont-à-Mousson Foundry) and also their famous cast-iron pipes. Everitube Situbé, branch of PAM since 1933, became EVERITUBE. The new business name remained until the 1985/86, when J.M. Coulon, General Manager, decided to revert to the former name of Everite. The latter remained in use until the death of the fibre-cement branch in 1997.

04-003. Everitube logo, Branch of Saint-Gobain Pont-à-Mousson. France. Private archives.

In 1953/54, winter was exceptionally cold in France and in Europe. Poverty was still very frequent. A priest, known as "Father Peter" *(l'Abbé Pierre in French)* made a hopeless call on a private radio telling that "*nobody could die of cold in the country*". The call generated a huge solidarity movement. Accommodation camps, partly made of asbestos-cement, were built in a very short time. The persistence of this priest made him very famous and beloved in the whole country *(his name is still often referred to as an example of charity in our 21st century).* In addition to his call, he searched industrials liable to help him to build prefabricated houses both cheap and in a reliable quality. Workers from Eternit heard his call and decided to offer one hour of their work. The managers of the company joined the workers' offer and on 9th December 1954, Father Peter was welcome at the Prouvy Factory, Hauts de France. In its issue no. 36, "*la Semaine du Nord*" (local newspaper), devoted a large article to the event which permitted to build seventy small houses near Paris. *"The Family Camp from Noisy-le-Grand received more than 252 families of workers and homeless, coming from the entire "Ile de France"* (large area surrounding Paris) *and even from other regions, who settled in tents of the US Army and igloos made of asbestos-cement"*[64] *shelters for homeless, near Paris.*

04-004 /005. Fibre-cement accommodation. Courtesy Musée d'Anzin

The same year 1954, Eternit created "S.E.R.T." *Société Eternit de Recherches Techniques*" (Eternit Technic Research Society) with the aim to confirm its will to be involved in basic research.

Creations of new companies and new factories were multiplying everywhere.

In November 1955, in Wanne-Eickel near Düsseldorf, North Rhine-Westphalia, Germany, appeared "Société Azet", which four years later turned into "*Wanit Gesellschaft für Azbest Zement Erzeugnisse*"[65] shared 50/50 between the French PAM and the German Thyssen. Commonly named "Wanit", it produced sheets and pipes. Let us note that as soon as 1980, it was in the forefront regarding industrial tests for production of fibre-cement without asbestos. A few years before its disappearance it was purchased by Eternit and soon after was near to come back into PAM' group, but this new project finally failed at the last moment. In between the town had also turned from Wanne-Eickel to Herne.

---

[64] Catalogue of an exhibition organised in 1995 by Theophile Jouglet d'Anzin Museum, with the idea of making neighbours know about the former industries of the region, among which Eternit.

[65] This Franco-German company was one of those which I had the honour to visit so many times till its death in 1987.

In 1956, the French Eternit built a new factory in Caronte, Provence, France where was installed a powerful pipe machine able to make 6-metre-long pipes (236in). Still in the same year 1955, in England, "Cape Board" registered a new brand: Asbestolux. In France, Everitube acquired the licence soon after. This new product, derived from "Marinite" registered by the American John's Manville, was made on Hatschek machines, same as asbestos-cement sheets. It was a blend of 20% amosite *(asbestos belonging to the amphibole group)*, 70% of a binder compound, in more or less equal parts of lime and silica. The remaining 10% was white cement, meant to help unmoulding after 24 hours and piling on special steel templates for air-steaming. The high rank of amosite permitted. a 0.8 density, almost half that of asbestos-cement sheets[66]. The new product met a very great success in false ceilings, dividing walls and even fire-doors

In Argentina, Fortalit and Monolit, the two main asbestos-cement makers of the country, merged and gave birth to Monofort[67].

In Venezuela, Canacit existed for already several years, but suffered of being located inside the town of Caracas and of big strategic errors. The two defects shortened its life and it did not survive long, though it revived under protection of the French Everite as we will see in chapter VIII "Countries overview".

---

[66] Details about the components of Asbestolux are due to a former Manager of R&D from the French Everite.

[67] My search in view of knowing more about it, unfortunately remained unsuccessful.

In France, still that same year 1955, Fibrociment de Poissy, already handled by Eternit for a good while, left its historical town of Poissy, Ile de France, and moved to the nearby Triel. Fibrociment de Poissy took advantage of the removal to develop its specialism, decorative inner moulding about which we will talk in the chapter VI dedicated to "Fibre(s)-cement, Art and Architecture".

In 1957, Eternit France in turn took advantage of its collaboration with Fibrociment de Poissy to leave the town of Poissy and moved to nearby Triel (*both towns near west from Paris*). New asbestos-cement products such as pieces of furniture for kitchen and bathroom as well as cool-box met a wide success, especially in butchers and delicatessen shops. In the 1950s, PAM did not run any enterprise in the Middle-East. Consequently, one of its managers started negotiations with the Pahlavi Foundation, which depended directly from the Shah.

04-006. Asbestos-cement cool box

They gave birth to Iranit in 1957. We will also have the opportunity to revert to this company in Chapter VIII.

In 1958, PAM modernised its factory of Dammarie-les-Lys, which lately had changed its name from Everite Sitube to Everitube. At the same time, PAM and the French Eternit joined their efforts in view of creating a new manufacturing process. In that purpose, they founded what they called: *"Syndicat d'Etudes de la Machine à Former"* (more or less: "Forming Machine Association Office"). A prototype was built in Dammarie-les-Lys. It turned into a total mess, justifying a fast withdrawal of the project from both partners.

In PAM, until the 1960s, search regarding asbestos-cement belonged to each factory which was free to act on its own behalf. In 1960, PAM created "SERAC" *(Service d'Etudes et de Recherches de l'Amiante-Ciment, or Asbestos-cement R & D)* in Dammarie-les-Lys. After a few years and several reforms, the office was linked to Everitube and named "*Usine Pilote*", *(meaning Trial Factory)*. Members of the Company would informally refer to as U.P. Besides usual tools common to every lab, two laboratory asbestos-cement machines, width 0.20m and 0.60m (8 and 24in) were settled in the new R & D office. They permitted to make simulations and all kind of tests, very near to future products, without disturbing industrial manufacturing.

04-007. Asbestos-cement Prefabricate class rooms from architects: Reubsaets, Thibaut and Gilles Web Eternit.

In the early 60s, France had to face both the "baby-boom" of the post war time and the return of millions of people back from newly independent Algeria. Such situation created a great and urgent need for schools and class-rooms which generated research for small and easy to move structures. Three architects, Messrs Reubsaets, Thibaut and Gilles, conceived a fast build structure, named R.T.G. from their initial letters.

At the time, asbestos-cement qualities made it a commonly used material in every type of building. There was logically no reason to avoid asbestos-cement, or fibre-cement, in the construction of such light and cheap buildings which contained between 10 to 15% of asbestos. Plenty of class-rooms were thus built in order to face up to the flood of pupils. The concomitant arrival of polystyrene foam boosted the use of asbestos-cement flat sheets as they were used in prefabricated panels, very often part of above described classrooms. At that time, such panels had no structural role, unlike what we will see later on. In Europe, it seems that the system was quite soon forgotten in favour of other technics considered as better. The idea took another look in America years later, especially in Mexico as we will see when talking about this Latin-American country.

In 1961, the French Eternit took a financial interest in SICOAC, Tunisia[68].

In 1962, the Asbestos mine of Canari, Corsica, already mentioned was the biggest enterprise in the island. The French Eternit created a new factory in Saint-Grégoire, near Rennes, Brittany.

In 1963, an asbestos-cement factory was settled near Saigon *(now Ho Chi Minh City)* Vietnam, and another one appeared, named ARAMIS in Eastern Pakistan, *(now Bangladesh).*

The same year, as Kurt Hünerberg already mentioned, the world totalised 173 asbestos-factories.

In 1964, Everitube built a new factory in Descartes, central France.

The same year, the Swiss Eternit newly had his own pavilion in the National Helvetic Confederation Fair. In the same time, Mr Roger Martin, now Chairman of PAM, surrounded by an impressive V.I.P. assembly, inaugurated with splendour the pipe factory of Chihuahua (North Mexico). A memorable blow-out took place on this opportunity[69].

Still in 1964, in Costa-Rica was RICALIT founded which, in the following years, turned to be an essential player in the fibre(s)-cement industry in Latin America. This point will also be developed in chapter VIII.

---

68 Catalogue of an exhibition organised in 1995 by Theophile Jouglet d'Anzin Museum, with the idea of making neighbours know about the former industries of the region, among which Eternit.

69 Roger Martin, "Patron de droit divin", page 216.

One must also keep in mind that the many new factories, in France, in Europe and other continents, were giving work and delight to consulting engineers, machine builders, *(most times German, Italian, or Swiss)*, felts and metal wires manufacturers[70], hoists manufacturers, transport companies of all kinds by road, sea or air, without forgetting bankers, supplying the indispensable credit lines. If we want to sum-up, let us say that the fabulous development of the industry we are talking about, had its share in what we use to call: "the 30 glorious years".
The following decade 1960/1970 went on with creations everywhere.
Eternit opened its Greek branch Helenit.
PAM built a pipe plant in Andancette, south of Lyon, France, on a site which had been chosen to produce mustard gas in 1918, but was forgotten as soon as war was over. When starting, this new plant counted only the 3 metres pipe machine (108in) equipped with a double feeding system shown chapter III, picture 03-034.
1967 saw the opening of the Brazilian asbestos mine of Cana Brava which had been discovered a few years earlier. We will talk about when coming to Brazil.
Still at the same time, PAM undertook to create its Spanish branch, Iberit S.A. in Valladolid, Spain. This later will not make it easy for PAM and its president in the following years. On the case we will also revert when talking of Spain.
In Senegal, President Shengor inaugurated a factory counting with two sheet machines in the town of Sebikotane, near Dakar.
The companies located abroad were seldom the fact of a single industrial. An association of asbestos-cement makers, called "TEAM" was in place. Under the leadership of Eternit, it was joining groups such as the American John's Manville, the British Turner Asbestos from Manchester, the numerous European Eternit and the French Everitube belonging to PAM[71].Most times, new companies resulted from joint-ventures, occasionally with shares belonging to the state of creation. On the other hand, cheap manpower in these countries avoided the need of expensive mechanism, contrarily to what happened in Europe. The industrial who had generated the new entity would usually stay in charge of administration and management, at least during the first years.

---

[70] Among which is also the author of the present book.
[71] Odette Hardy Hemery, "Eternit et l'Amiante 1922-2000".

In 1967, the French Saint-Gobain, associated with the American CertainTeed[72], in view to prove the dynamism of the French group and its will to penetrate the "New World"[73].

In 1968, J.P. Guerber, in charge of R&D for asbestos-cement in the research centre of PAM, wrote a manual for trainees where one can find: technics, professional vocabulary in several languages, machine diagrams and process of the time.[74]

On the same year, in Saudi-Arabia, appeared Amiantit in Dammam, Ach-Charqiya Province.

In 1969, the French Everitube added a 5-metre pipe machine to its factory of Andancette, in order to support the already existing 3 metre one.

04-008. Everite Andancette 5 metres large pipe Machine 1969. Private archives

[72] Roger Martin, "Patron de droit divin".

[73] Details will be developed in chapter VIII.

[74] Lot of technical

It might be worthy of interesting us to some details of the inside life of a factory. The short following chart can give us an idea about pleasures and griefs felt by workers according to the amount of the quarterly bonus. It is obvious that the amount was directly linked to the gross operating result of the plant that means: orders received, energy cost, work-power cost, and also good or poor management.

1987: How quarterly bonus must be calculated.					
Amount	***1st***	***2nd***	***3rd***	***4th***	***G.O.R. yearly aim***
100F	*22 M\F*	*27 M\F*	*32 M\F*	*37 M\F*	*118 M\F*
200F	*24 M\F*	*29 M\F*	*34 M\F*	*39 M\F*	*126 M\F*
300F	*26 M\F*	*31 M\F*	*36 M\F*	*41 M\F*	*134 M\F*

"Each worker will get in cash Francs 300 x 4 = Francs 1200 in the year. Bonus is equal for everyone, whatever his level in the company is. G.O.R.= Gross Operating Result".

In 1970, an event occurred on the French industrial activity. PAM, king of ductile iron pipes, and Saint-Gobain number one in the glass industry, merged, giving birth to a worldwide industrial giant. Everitube and its already existing foreign branches were members of the new company. This event highly impacted the development of the asbestos-cement industry under French influence. We will have the opportunity to observe it in the next chapters.

This same year 1970, a Frenchman who managed the German branch Wanit, already mentioned, had a house in Haute-Savoie, French Alpes famous skiing region. He or dered his roofers in charge of restoring the roof, to set asbestos corrugated sheets on the framework before placing the tiles. The roofers laughed at him. When the snow melted the following spring, the idea soon appeared a good one. Water ran down towards the gutters viathe canals of the sheets, without filtering through, and did not stain the ceilings. We ignore if this gentleman invented the system, later on named "canalit", but in fact it was a true success. It is based on an old technic widely used by the ancient Greeks and Romans, known as "tegula-embrix", which consisted, and still consists, in covering one edge of each tile by the canal shaped edge of the other one. This method, applied to corrugated sheets was very successful in the 1970s/1990s. At Everitube, two types were developed, "Atlantic" and "Mediterra- nean ", each one being adapted to size and shape of the tiles of the region., which equally knew a wide success in countries with traditional tile roofing.

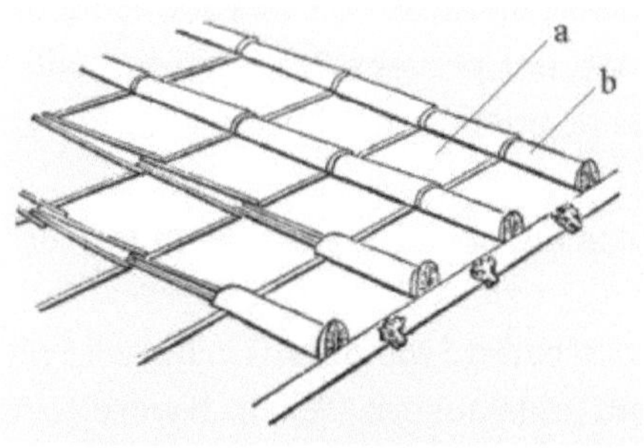

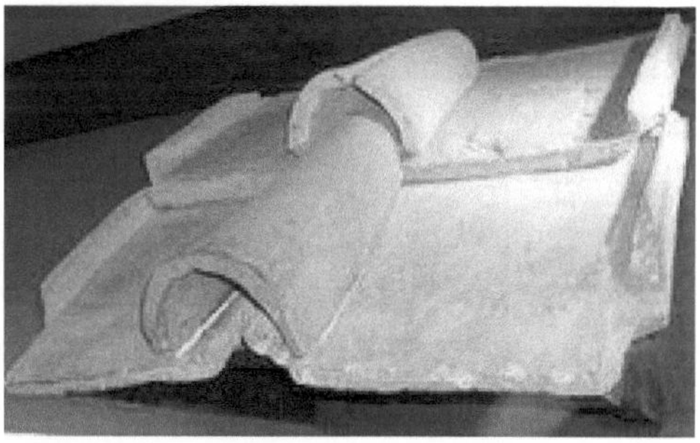

04-009-010. Tegula-Imbrex system imagined by ancient Greeks and Romans. Wikipedia

At the same time, Eternit also created a new product with the same purpose. All these products had the big advantage of offering a perfect waterproofness, re-use of old tiles, and respect of heritage as they were invisible once the work was completed.

In 1970, France was beginning to face the asbestos problem and health hazards with workers in close contact with the fibre. Eternit participated to the birth of AFA *(French Asbestos Association)* and soon after to that of AIA *(International Asbestos Association)* organisations gathering the main industries using asbestos as a raw material. For our record, to be listed among these latter: the building industry, the automobile industry, shipbuilding industry and so on. Every two years, AIA organised international conferences. The CEO of Everitube was president for a while. The defended theory was the so called "asbestos use control", according to which under a defined threshold of fibres in the air, there was not any risk. At that time, we were still far from the 1996 INSERM[75] report which was going to inflame the matter. In spite of the danger, the 1970 decade carried on with the same line as the previous one.

In Descartes, the recently build Everitube factory, inaugurated a social centre in July. Soon after the merger between PAM and Saint-Gobain, the general management decided that pressure and sewerage pipes should be sold through PAM and the only pipes bound to the building industry (*down lines from gutters, rubbish chute, etc)* should

[75] INSERM: Institut National de la Santé et de la Recherche Médicale or *(French)* National Institute for Health and Medical Research.

be sold by Everitube. At Everitube, production managers remained convinced that PAM's salesmen were favouring their own ductile pipes, which was reducing the sales of asbestos-cement pipes. The situation did not prevent Everitube to build a pipe factory on the bank of the river Seine in Saint-Etienne du Rouvray near Rouen, Normandy, France.

At the same period Eternit was also building a pipe factory in Terssac, near Albi, Tarn, southwest of France.

In 1971, George and Bernhard Alsptaeg brothers created Swisspor AG in Boswil Switzerland, as well as two productions units, Kork AG and Baukork AG, which were going to become a big group with a worldwide future.

In 1972, both to reduce costs resulting from imports from Germany, and influenced by the French government, then careful to limit the social problems resulting from coal mines closure, Mr Lucaussy, importer of asbestos cement products, joined Everitube *(Subsidiary of Saint-Gobain)* and created a factory in Saint-Eloy -les-Mines, Puy de Dôme, Central France[76]. The new factory counted one pipe machine and one sheet machine and gave 290 people work. Its products were sold under the name of Franconit[77].Soon after, Everitube who had just been an investor turned into majority shareholder and the name Franconit disappeared. The first manager of the factory, a tall and smart guy, who had learned the work in Bassens factory, did not keep his job for a long time as he soon became a victim of asbestos. His superior, who had him as a trainee a few years before and who kept him in high esteem, visited him in the Paris hospital where he was in intensive care. At the door of the room, his spouse suggested to the visitor to avoid the shocking sight. Years later when I met the retired manager within the scope of the present book, the memory of the visit to the hospital still left marks.

In 1973, Saint-Gobain actually entered the US building market, by increasing his shares in CertainTeed. Following his traditions, the big French group became majority shareholder as soon as 1976, and unique owner in 1988[78].

This very same year, began in U.P. *(Usine Pilote ,that is: test plant)* of Everitube searches to discover fibres liable to replace asbestos.

---

[76] The place was very near from the land of Valery Giscard d'Estaing, future president of France.

[77] Origin: PCM *(Ponts et Chaussées et Mines = Civil Engineering and Mines)* March 1972.

[78] Origin: François Malye: « Amiante 100,000 morts à venir » (Asbestos 100.000 death to come).

The first oil shock took place in 1973. In Europe, the tendency was to reduce the consumption of the newly valuable fluid. Let us remember the French advertising slogan: "*la chasse au gaspi*" (informal French for waste hunting). In the contrary, oil producing countries began to gather profits. Such was the case of Algeria which, thanks to its huge gas reserves, perceived comfortable and welcome unexpected incomes. The country acquired, among others, three asbestos-cement factories, via a consortium set with Société Générale de Belgique *(the biggest ever bank in Belgium)* in Brussels. The three factories were exactly identical, buildings, machines and organisation. One was located in Bordj-Bou-Arreridj, commonly called BBA, east coast, near Constantine, one in Meftah near Alger, and the last one in Zahana, west coast, near Oran. Each counted one sheet machine, supplied, settled and started by Eternit and one 5 metre pipe machine set in the same conditions by Everitube. The French engineers in charge of the starting assistance were somewhat jealous of their Algerian colleagues, who were doted of the most modern machinery of the time[79]. They would have like to get the same ones in France.

In 1974, Eternit produced 800,000 tons of asbestos-cement in its French factories[80].

In the town of Lahti, Finland, a modest asbestos-cement company was born, which soon specialised in frontage cladding and became a renowned exporter worldwide, mostly in Northern Europe and Russia, as well as in the Middle-East.

In 1974 appeared Eternit Sapele, founded in Nigeria by Eternit Belgium.

In 1975, the French Everitube won a new contract concerning Cuba, including 3 asbestos-cement machines to build under Castro's government. This contract generated some funny incidents which I will tell you in chapter VIII dedicated to countries overview, where we will escape from pure chronology.

In the years immediately following, the rhythm of creation tended to reduce. One must admit that assimilating and managing all the new companies was rather a heavy job. Simultaneously, doubts were growing regarding the future of asbestos.

---

[79] I had the opportunity to visit these factories twice in the scope of felts supply. I could note the modernism of the newly delivered machines, but on my second visit, some 18 months later, I was very disappointed by the lack of care in maintenance. Let us note that when the French executives left, once the assistance contract was over, the management was entrusted to "political leaders" and not to the people who had been trained in France and Belgium in view of the task. No need to say that the politician had only little competence to do the job.

[80] Origin: Jouglet d'Anzin Museum, already mentioned.

Its health damages began to be actually well known. R&D departments were working hard to discover asbestos replacement solutions. The miracle fibre was becoming the damned one.
A discreet, though important, event occurred in 1976. Stephan Schmidheiny became general manager of Eternit Switzerland, which he deeply modified in the following years and whose history would justify by itself an entire book. Stephan Schmidheiny created "World Business Council Sustainable Development Group" WBCSD[81]. We will also revert to this case in the countries overview chapter.
Simultaneously, the asbestos cement pipe market began to decline in Europe. Eternit France closed its recent factory of Caronte near Marseille, as soon as 1979. Not long before, it had counted up to 600 workers.
The French government was beginning to become aware of environmental problems. An agreement was built between government and asbestos cement manufacturers considered as polluters. The agreement known as "zero emission" was signed ceremoniously on 17th May 1980 with Michel d'Ornano then "Environment and Life Quality Minister". Companies committed to proceed with some trial acts such as recycling water and sludge. In exchange companies received subsidies from the ministry.
In most countries, industrial water, often with a high Ph resulting from the nature of cement, was quite always rejected to the nearest big or small river. Risks for water quality and wildlife were also high. The asbestos cement industry was from the beginning a great consumer of water and cement and did not escape to the case. It was then logical to tend towards "zero emission", which required heavy investments, justifying financial commitment from the state. The phenomenon was still increased by the arrival of pulp in the manufacturing process, and also harmful to organic life. Every time a risk appeared; local angler's associations paid very close attention to the respect of applicable standards.
Everitube, the Saint-Etienne factory, near Rouen, Normandy, was chosen for the test in spite of its already foreseen closure to come soon. For once, let us be happy that French public and private managers were considered as pioneers.

---

81 "WBCSD" is an organisation counting more than 200 first rate companies, managed by CEOs working together in order to shorten the transition period towards a sustainable world.

At the same time, Stephan Schmidtheiny, President of Eternit Switzerland, officially informed that the Group he was managing was definitely leaving the use of asbestos. He also declared that it is intended to turn the group into the worldwide leader of asbestos elimination as industrial material.
Let us read an extract of what he told us in his memory book named:

My Path-My Perspective

"The controversy over the potentially harmful effects of asbestos dust was a shock to me in many respects. I myself had been exposed to asbestos fibres during my training period in Brazil. I frequently helped to load asbestos bags and pour the fibres into the mixer, breathing in deeply the whole time due to the exertion the work entailed. At the end of a hard day's work, I would often be covered in white dust. "This taught me that important people do not always say important things, and that oftentimes better results can be obtained by going beyond established rules."
I could not determine by myself the actual risk level involved in the manufacture of asbestos-cement products. Our advisors believed that the scientific studies purporting to establish the harmful effects of asbestos were rife with contradictions. I personally felt that the lack of a clear scientific and technical consensus on asbestos and the inherent unpredictability of its effects rendered impossible any reliable planning and risk evaluation. And aside from being worried about risks to the health of employees of group companies, I reached the conclusion that this was not a very promising business in which to be".

In 1982, the ex-Franconit, created just ten years earlier, was definitely closed by Everitube. Things were somewhat different in other countries and other continents. For instance, the Spaniard Uralita, on that same year, opened a branch in Riobamba, Ecuador, sold a pipe machine in Egypt and a sheet machine in Tunisia. Each sale necessitated supply of technology and assistance of specialists, at least for the first operating months.
In the same year 1982, CPA *(Comité Permanent de l'Amiante or Permanent Asbestos Committee)* was created in France. It was constituted of independent members, from very

various origin[82]. On this opportunity, we must remember that asbestos-cement industry was not the only one to use this fibre in big quantities; motorcar, building construction and shipyard industries were also giant consumers.

During the same period, Everitube developed another variety of Asbestolux, and renamed it "Everifeu" *(which could possibly be translated as "Everifire", though both in French and in my translation trial, I am not convinced of the relevance of the words*). The interest of the variety was the absence of asbestos. The latter was replaced by pulp without reducing, neither the thermic isolation quality nor, more surprising, the fire resistance ability of the product. Everifeu was registered as a trademark in 1984, and was used in quite a lot of building works. A bigger development would have required large investments at steam-cure level, which shareholders refused. For this reason, the product did not meet the wide success that it was deserved[83].

In 1984, Eternit Switzerland was divided in two entities: Asbestos-cement was attributed to Stephan Schmidheiny, and Cement was attributed to his brother Thomas.

In 1985, Everitube, started a new slate machine, called HX1, in its Descartes factory, Central France. This new machine produced by Bell Engineering AG, Kriens / Lucerne, Switzerland, was equipped with 12,000 tons (13,200 US tons) Simpelkampf stamping press. This press was in charge of straining the slates, previously cut and piled on steel templates. The machine also included a very complex system of running carriages, which were a true challenge for the new plant manager, for the people of maintenance and for those of the "New Works Department" of the group. The pace of arrival of sheets from the making roller, turned the smallest incident into a complete stop of the whole line, with all that it implied...

---

[82]Origin: Jouglet d'Anzin Museum, already mentioned.

[83] As a former supplier of the fibre-cement industry, I can add that the use of pulp instead of asbestos seriously reduced the lifetime of cutting discs, which is very costly due to the high quality of the steel required by such discs. Years later, the problem was solved by the replacement of cutting disks by high pressure water jet streams.

IV Worldwide outburst

Putting this HX I machine into operation on September 18th 1985 was the occasion of a laudatory report in the group inner journal.
In the following years "asbestos-less sheets" began to be manufactured. Asbestos was replaced by a blend of cellulose and synthetic fibres. This new process required the moving and updating of the entire pulp treatment machinery from Dammarie-les-Lys to Descartes.
A short while later the sales department obtained an outstanding order. The Disneyland Park Paris was to open soon and the roof of the Disney Hotel needed pink coloured slates. Descartes HXI was going to produce these slates, quite a nice shot for the new "asbestos-less slates" and sweat for the people who had to develop and produce them. The whole team of Everite was talking about this exciting event. Everyone expected a great effect on future sales as the running huge Disney project was on the front page of every French newspaper.
Millions of visitors from France and abroad were expected and the amount of investments needed by the park impressed everybody. A high-speed train station was going to be built nearby in order to help visitors to reach the place. At Everite, the purchasing department pretended to gain rebates from suppliers as they were supposed to benefit of the aura. Consequently, they could not refuse to reduce their prices as the company was going to make them more famous ...[84]

[84] Private memory of the author who, at that time, was felt supplier at Everite.

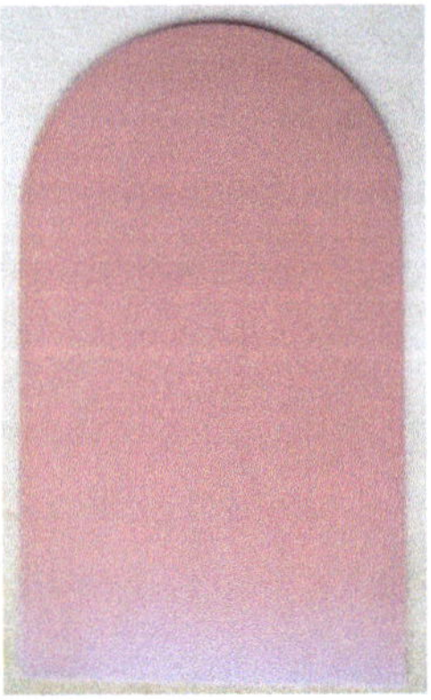

04-019 Asbestos free slates, coloured on request of Disneyland. Private archives and 04-020. Disneyland Hotel in the famous parc near Paris. Private archives

In spite of the achievement of this masterstroke, we will have the opportunity to see that the factory which had received large investments, did not survive very long as the arriving crisis was brewing for a long time. As soon as 1981, the Australian James Hardie, was involved in fibre-cement pipes with less asbestos. In 1985, he gave up asbestos in its products intended for building industry and did the same with pipes in 1988.

While some factories benefited of important investments, such as we could see with Everite Descartes. Others such as Everite Bassens, long the emblem of Everite, were closing. Nowadays, they still remain as industrial wasteland.

The case of Bassens Factory was highly representative. Closed by decision of the head office of the group in 1987, this factory created in 1917 had been the model and executive's incubator for the entire company. It counted 150 workers in 1922, 500 in 1939, 1160 in 1955, and still 280 when it was closed.

One of the former workers told me in 2016: *"In the lucky years, when there was more work than we could do, sometimes I was proposed to do a second day work just after the first one. The second day was needed due to unexpected problems. For instance, a big operating incident on one of the machines required hours of cleaning as the cement paste had spurted all the way around. They did not count our time when the shift was due from 6 p.m. to 4 a.m. but the work was already finished at 11 p.m. Then I could go to sleep somewhere on the shop the system, happened to double his week pay but the work was already finished at 11 p.m. Then I could go to sleep somewhere on the shop the system, happened to double his week pay*

This old man, meanwhile passed away, had retired more than 30 years ago. During the years, he had in mind to make something special to celebrate the coming millennium. And he did it! From memory he constructed a model sheet machine, including every detail of the process from the storing bins to the corrugating device. He did not forget neither the desk nor the worker fitted with a high-pressure water lance to clean the fresh paste wastes. He even included the lift truck carrying the sheets to the storage area. Then he put the model on a small trailer and showed his work through the towns around the closed factory[85].He won the respect of most of his former working colleagues and laudatory articles in regional newspapers. Didn't he deserve it?

---

85 When he showed me the model, I feared that it be lost after his death. I invited the Musée des Arts et Métiers *(Museum of Arts and Craft)* Paris to take care of the model. They did not even answer. Then the archivists of Saint Gobain were interested in the idea, but they changed their mind. Finally, "archivists of Bordeaux Metropole" wholeheartedly accepted to receive it and to exhibit it for future visitors. Thanks to them.

04-020 Hatschek machine. Model due to Marcel Descombe, former workman at Everitube Bassens factory Pictures by ©B. Rakotomanga, Archives Bordeaux Metropole

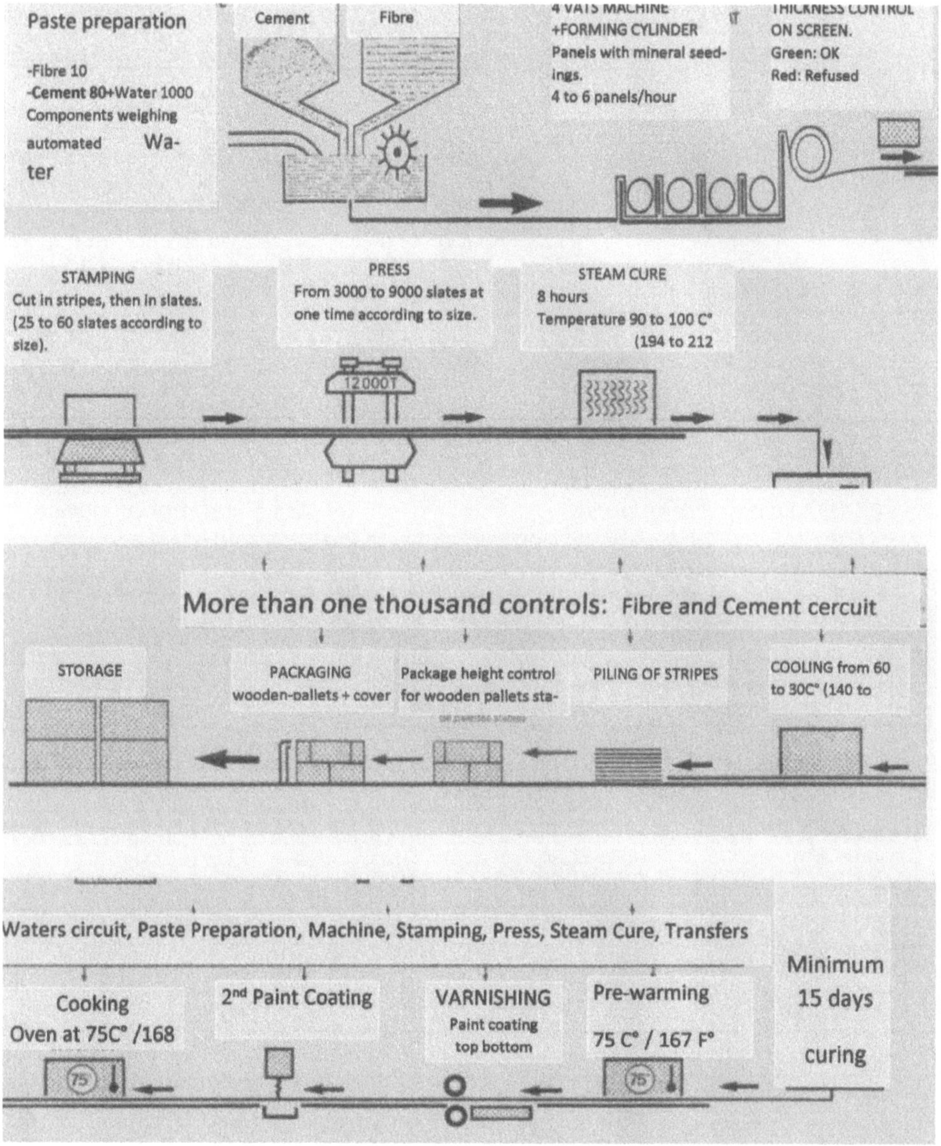

04-021 New Bell Machine process in Everite Descartes Factory. *Pierre Picavet Archives*

04-022 Siempelkamp press

04-023 Sheet motion device

04-024 Other sight on the sheet motion device

04-025 Suction pad for sheet motion.

NEWS OF THE GROUP

IMPLEMENTATION OF A NEW LINE OF FIBRE-CEMENT SLATES AT EVERITUBE-DESCARTES FACTORY

On September 18th, Everitube implemented a new line for the manufacturing of fibre-cement ARMOR S slates in its factory of Descartes, Indre et Loire, France. The new line has a capacity of 40 000 tons corresponding to 2 million square metres[1] of slate roofing, i.e.12% of the national market.

Technically, Engineering of this line, which corresponds to a 70 million French Francs investment, was entirely conceived by Everitube with help of the best European machinery makers.

Being the most modern in Europe in its category, the line is 100% automated, from paste preparation to placing on wooden pallets, which clearly allows to improve productivity and to offer first quality and competing products to areas with traditional slate roofs (Britany and Loire-Country).

On a social level, implementation of this new line required a global training plan. 8000 hours (900 000 FF), were dedicated to workers training. Thanks, to the introduction, for the first time, of quality improvement teams in the training plan and to the creation of suggest circles, the training course to the new line, entirely lead on site, was very personalized. Regarding environmental protection, thanks to original technical solutions same as in every Everitube factory, the new line participates to the environmental protection plan signed with the "Environmental Ministry".

The implementation of the ultramodern new line of fibre-cement slates, which fits the global plan bound to improve Everitube productivity, confirms the competitiveness of Descartes factory in the field of roofing materials.

President Jean-Michel COULON inaugurated the new line in the presence of Messrs Roger FAUROUX, Jean Louis BEFFA, Alain de METZ, of local authorities and of some customers of the Company.

04-013 Private Archives.

[1] 40000 metric-tons =44092 US t.
2million square meters ≈ 2.4 million square yards.

In spite of left and right developments, creation, constant technic progress, everything was not pink in the asbestos-cement world. From time to time, a corrugated sheet would break under the steps of a roofer generating a more or less serious accident. The manufacturer was sometimes taken to court, then research was made in view to strengthen the sheet endurance. Trials were done to incorporate discontinuous reinforcement inside corrugated sheets by including strips of synthetic ribbon between two main layers right behind the making roller. The foreign body had to be incorporated more or less in the centre of the thickness. The operation appeared rather delicate...

At Eternit, a polypropylene ribbon was adopted and a "ribbon throwing device" was conceived. An agreement was signed between Eternit and Everite to allow them to use the process. A former Everite engineer, retired for a long time, told me how he participated to provide the system at Everite after the deal had been signed between the two companies:

"Indeed, I was involved in the trials to reinforce the waves of the AC sheets with polypropylene ribbons. According to the process conceived by Eternit; it consisted in setting the ribbon, at mid-thickness, in the bottom of each wave while the sheet was being formed.

A system of "ribbon guides" was located behind the making roll, in line with the bottom of the future wave. When the sheet was forming, we would count the rounds of the making roll since the last cut. Then we would 'send' simultaneously the five ribbons, (one per wave) previously cut at the length of the sheet, between the 4th and 5th round. The ribbons were then grabbed between the felt and the roll which was already covered by the 3 first layers of paste".

I settled this device, as experiment, on one of Dammarie's Hatcheck sheet machines. It did work, but the usefulness of the reinforcement was uncertain: indeed, when sheets were set, nothing would link the polypropylene ribbons to the frame of the roof. In case of break, under the steps of the roofer, pieces of sheets, ribbons and... roofer would joyfully tumble down. Truly: Unreal safety!"

I settled this device, as experiment, on one of Dammarie's Hatcheck sheet machines.

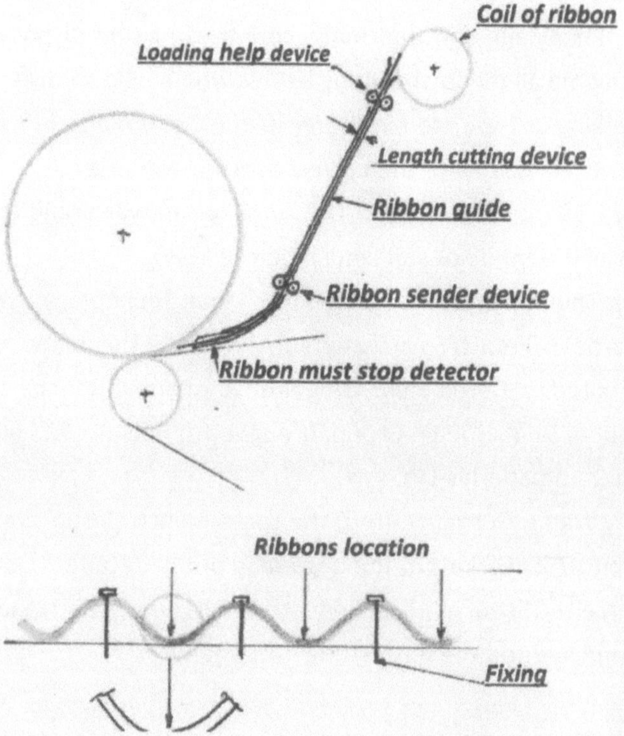

04-026 Memory hand drawing by Robert Bessiron Retired Engineer from Everitube

At the same time, among manufacturers, the word "asbestos-cement" was becoming forbidden. The long-time popular denomination "fibrociment" was replaced by the more neutral one: fibre-cement" as everyone was preparing the replacement of the now damnable asbestos by other fibres. Thus, the title of the book you are now reading, which bears a big S at the end of the word fibre. Indeed, research for replacement fibres were speeding up. Everyone was hiding from the other, perhaps to avoid to see his fears regarding the future, but overall, with the great hope to be the first one to discover the miracle fibre which would grant conversion, with a new start and a rich future. Every investor knew that world population was increasing, that buildings were taking age and that many countries had no disposal of natural raw material to welcome all the forthcoming babies who would soon require dwellings... R&Ds were working at top speed, not only in Europe, but also in America, Australia and even in South Africa one of the biggest asbestos exporter.

As soon as 1974/1975, in Everitube, "U.P." *(trial plant)* was urged to find a solution, though industrial trials did not start much before 1980.

In 1984, Eternit Thiant factory was prized for "Clean Technology" from the hands of Messrs Bouchardeau, French Environmental Minister[86]. The following year, the same price was attributed to Eternit Saint-Gregoire factory near Rennes, Britany, France.

In 1987, Everitube, Saint-Gobain Group, brought a standstill of its old pipe machines in Bassens and Dammarie-les-Lys.

The same year, Wanit, Germany, from the same French Group, was entirely closed after having up to 750 workers at the beginning of the decade.

We could enlarge the list of all those industrial sites, which had made the happiness of every one, and later on the misfortune of so many.

"The very story of apotheosis, alas is to lead to decline".

Roger Martin, Patron de droit divin.

[86] Amiante 100,000 Morts Annoncés, by Francis Malye, already mentioned.,

V The crisis

"Trials by hundreds. Complaints by thousands. The worst that every organiser of controlled asbestos could fear".

François Malye[87]

A crisis doesn't happen all at once like a car crash or the sinking of a liner. No, it's rather like an inner ache, a small pain somewhere in one's body, a pain that one feels but refuses to mind about, convinced that "it will go as it came".

But over the months and years, the pain intensifies and the question cannot be answered. First by oneself and finally when meeting friends, confidantes or colleagues. At the beginning questions remain discreet, then more precise:

"- I had a call from the occupational health doc, he talked about several similar cases among the guys of the store-shop...and you? You got troubles?

"- Just a few by now, but I met X... a while ago. In his country, authorities seem to be preparing rules to supervise the use of ...".

"- Me too, during my recent trip to Germany I met some colleagues, they are working hard searching replacement fibres..."

"- We too, you know what? I just contracted two chemists that I poached from a Swiss company making synthetic fibres. The worst must be foreseen... And costs will skyrocket!"

"- I fear some questions during my next central works council... I got a staff member who does his best "top... me off".

---

87 « Amiante : 100,000 MORTS ANNONCÉS » *(" Asbestos : 100,000 death foreseen ").* François Malye, Cherche Midi 2004.

We saw that ancient Romans remarked many deaths among young people who had woven asbestos tablecloth.

In a modern world, health hazards related to asbestos appeared as soon as 1899. In London, Doctor Murray observed an asbestos linked death this very year. That means even before the asbestos-cement industry started.
In 1906 in Caen, Normandy, France, a certain Auribault, regional work inspector[88] noted a connection between exposure to asbestos fibres and professional death in the news-bulletin of his organisation.
1931 in the USA and in UK, rules were enacted in relation to asbestos workers conditions.
On April 25th 1951, the Netherlands government declared asbestosis a professional illness and announced the "Silicosis Act" which described the corresponding risks.
Large North American and European groups with powerful lobbies managed to convince that there were so few cases, that they were not worse talking about... A few isolated cases were not going to paralyse the needs of the building industry, moreover when comparatively cheap, easy to manufacture and to use. It is quite easy to pass the buck to companies' managers or to politicians, but one cannot forget the huge demand, first in Europe and USA, then in the whole world.
Around 1973, the existence of asbestos cancer became apparent to everyone.
In April 1977, France published his first decree regarding the use of asbestos. It remained applicable till the publication of the following one in July 1983.
Indeed, the actual sanitary risk was not essential until the 1980s. In most cases, company research preceded governmental regulation, even if at the same time lobbies were very active near ministries and MPs. As soon as authorities tried to introduce restrictive regulations, there was blackmail in employment not only in relation to asbestos, but also in many other topics which we now consider as obvious.
In most countries, one observed the great administrative slowness to demand industrials to take the necessary steps to seriously protect workers, even when dangers were known for a long time. This can be explained by the power of lobbies, not only in the asbestos-cement industry, but also among all those using asbestos, automo-

[88] Regional work inspection was created in France as soon as 1886.

biles, buildings, ships etc. Let us remember the above described "asbestos use control" which claimed to respect its own norms. In fact, hazards were reduced to none. Quarrels between industrial representatives and State Departments, doctors, specialists and lawyers were endless. Courts were referred to, casualties died and families mourned, but nothing happened...Years elapsed and the situation got worse.
But are we also wondering how we would behave when we face the choice, whether we want to take a risk and kill an apparently flourishing activity with all the political insecurities implied by such a decision? The difficulty of the choice, perhaps in a coward way, prevents me from commenting on such a sensitive subject.
We already noted that the health issue implied to every industry using asbestos. Moreover, many products which neither you nor I could do without, met a similar situation. The evidence lies in the example offered by the oil industry about which an article in the French magazine "Le Point" n°2317, dated February 2nd 2017, signed by Hélène Vissière gives a clear idea:
"A two-year old inquiry revealed that the managers of Exxon Co. were aware, that fossil energies impacted global warming since 1977. At the beginning, they carefully hid the truth, then they lobbied like mad men to prevent establishment of the rules. Worse, they tried to create doubts by subsidising at least 43 climate change sceptic groups with millions of dollars, imitating the method of tobacco producers". Indeed, when forty years is a long time in regards to a man's life, it's really a short one in regards to the blue planet.
This is how in 1991," the Asbestos Institute "of Sherbrooke, Canada, well aware of the coming dangers, organised a wide meeting named: *"The International Conference on Asbestos Products"*. The idea was to defend "asbestos controlled usage" and in a certain way check the world opinion pulse on the subject.
Organisers thought that in Europe or North America the subject might have generated violent polemics. Consequently, they chose an Asian country that is less sensitive to asbestos problems from the outset and where the meeting should take place. Kuala-Lumpur, head town of Malaysia, was elected to welcome 208 guests proceeding from 34 countries.
By luck, one of the guests, formally in charge of R&D at the French Everite, found for us in his archives the communication that he presented to the members of the con-

ference and the detailed list of participants. Thanks to him we have a detailed analysis of the progress of the time in terms of replacement fibres in the products we are talking about.
Apart from Malaysian neighbour countries that have sent many delegates, we can note the number of representatives directed by distant ones.

- Canada: 31, 23 of them were sent by organisms such as states, mines, urban universities and by the own organiser, the Asbestos Institute.
- Europe: 23 among them:

    France: 3
    Germany: 1
    Switzerland: 3
    UK: 5

From other continents:

Brasil: 2
Mexico: 3
USA: 7
URSS: 5
Zimbabwe: 4

Large Extracts of the communication presented on November 3rd 1991, called: "Technical and economic aspects of substitutes vs asbestos in fibre-cement products":

"Since about 1975 a considerable amount of work has been carried out to find other suitable fibres to replace asbestos in asbestos-cement manufacture, or to develop new fibre reinforced cement composites and related production techniques able to replace a-c products directly as wells as existing equipment.
The purpose of this study is to provide a broad review of these attempts, which have led to a wide range of industrial development stages, from reasonable success to complete failure. Emphasis is put in a first part on the possibility of converting a-c technology to the use of other fibres, especially wood pulp, associated or not with man-made fibres. As a consequence of asbestos replacement, there is often a need to modify the cementitious binder.

Technical and economic aspects of asbestos substitution, as well as short term performances and durability of finished products are analysed.
Among the main families of a-c building products (flat sheets, roofing slates, profiled sheets) the most critical situation, should asbestos replacement occur, would undoubtedly concern manufacture of corrugated sheets, which would require high additional investment, and would result in a dramatic increase in cost, though not offering really improved performances. Special low-cost formulations, with limited machinery modifications have been tried, particularly in developing countries, but without noticeable success...
For nearly twenty years, when health hazards due to an uncontrolled use of asbestos were well known, many attempts have been made to propose substitutes to asbestos cement products using other fibres or other reinforcement materials.
At this stage, one may wonder whether asbestos-cement, born with the 20th century, is due to end with it?
Already, at a RILEM[89] Congress in London in 1975, H. KRENCHEL was asking, with some scepticism: "Can asbestos be completely replaced one day?" The question is still worth raising...
Industrial substitutes to asbestos-cement appeared for the first time in Europe, particularly in the U.K in the years around 1975. These products were first calcium silicate boards reinforced with cellulose and mica.
Then, alkali-resistant glass fibres, invented ten years earlier, were tried, alone or in combination with cellulose. Later on, polymer fibres, like polyvinyl alcohol (PVA) or polyacrylonitrile (PAN) were applied to cement reinforcement in the early 80's. Polypropylene which was already known for a long time for limited applications in concrete reinforcement, originated some innovative developments...

Brief summary on Hatschek process:

Hatschek process, schematically shown on figure 05-001, consists in continuously filtering on rotating sieves clothed with a fine wire mesh, a diluted slurry of asbestos fibres and Portland cement. The concentration of this slurry, which may contain fillers, pigments, various additives, is usually between 150 and 200 g/l of dry solids.

89 RILEM: International Union of Laboratories and Experts in Construction Materials, Systems and Structures

Three or four elementary layers are successively picked up and build the basic lamina which is transferred to an endless felt, dewatered and finally wound on the accumulator roll until the desired thickness is obtained. The "green" sheet is then cut, unwound and put on a steel mould for the time necessary to the curing of cement. At this stage the material may be mechanically corrugated to various profiles or hand moulded.

Besides the Hatschek machine, a complete production line includes equipment for handling and processing fibres, storing and weighing other raw materials, mixing the slurry, and a complex installation to recycle backwater from the sieves, water used to wash the sieves and the felt, and green waste which is dispersed and mixed with fresh slurry.

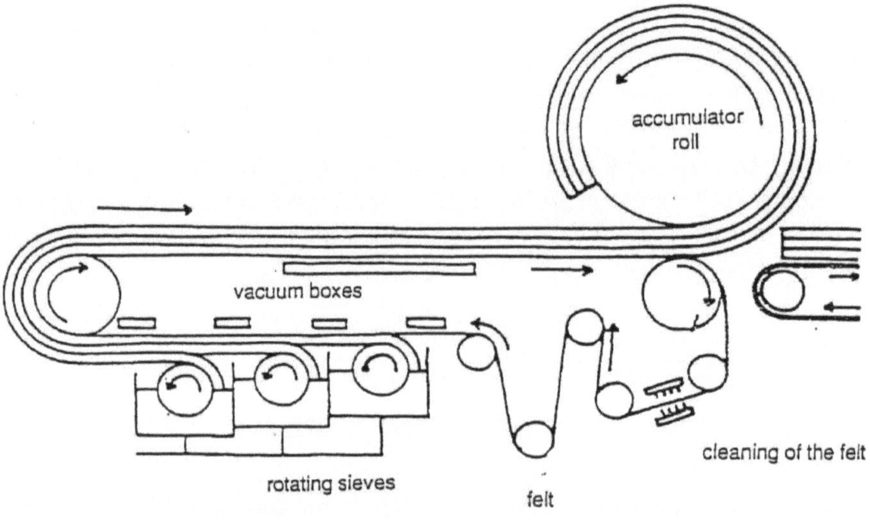

05-001 Hatschek corrugated sheet machine, corrugating device not shown

For about 10 years much attention has been paid to:

a) *improvement of health conditions for handling and processing of asbestos: asbestos bags are now introduced in an automatic bag opener under depression, empty bags are finely ground and this ground material is recycled in asbestos-cement slurry.*

b) *most complete recycling of process water, "sludge" from cleaning of the machinery or produced at machine stops, and rejected hardened products which are ground and recycled as a "fibrous filler".*

In most advanced production lines, the final target of "zero waste" is almost achieved

Manifold role of asbestos fibres in Hatschek process

We will only deal with chrysotile asbestos. A large number of properties of chrysotile are very useful in the manufacturing process and performances of asbestos-cement:

- *Hydrophilic character, allowing an easy dispersion into water.*
- *Stability at high pH: the pH of a cement paste is about 12.6 which is equivalent to a saturated lime solution.*
- *Surface properties: chrysotile has an isoelectric point at pH 12.6 which results in a natural flocculation of asbestos cement-suspensions.*
- *Structure in bundles of tiny microfibrils (diameter of the elementary fibre is about 20 mm) which imparts to asbestos a very high surface area (1 to 2 m^2/g) and a great flexibility ... from these properties derive the filtering capacity of asbestos and its role as a micro-reinforcement or crack arrestor in a cementitious binder (hence, asbestos-cement would be more rightly called a mg-Lahase-material, rather than a composite material as it usually is).*

  -High mechanical characteristics which enable asbestos to act as a true reinforcing material in the elastic field...Resistance to hydrothermal treatment: steam pressure curing of a blend of cement and silica at about 170/180^{0}C and 7 to 10 bars. - incombustibility...

It is well known that Hatschek process was derived from cardboard manufacturing process. One might then conclude that cellulose fibres are best suited to replace asbestos in this process. Moreover, cellulose has been used for a long time in combination with asbestos to produce lightweight, flexible flat sheets, with fairly good acoustic properties and good workability. During World War II, due to rupture of asbestos supply to Europe, cellulose cement products were manufactured in several countries, and it is likely that some of them are still in use....Mineral wools, from glass slag or rock, have also for long been added to asbestos, mainly as filtration aids. Some research work has been devoted to glass wool cement composites which may exhibit valuable short-term properties, but these materials are becoming completely brittle very quickly and are not acceptable....

Wood pulp is then appearing as the one fibre suitable to substitute asbestos in Hatschek process. However, we will see hereafter that many difficulties have to be overcome in order to reach this target, from a three-fold standpoint:

-process,
-durability
-cost.

Laboratory experience clearly shows that asbestos is favoured by the extreme fineness of its fibres (around a few microns, even less). In asbestos-cement, the efficiency of filtration lies within a narrow range, about 85 - 90 °/0, depending on the degree of opening of the fibres in the factory.

If asbestos is replaced by cellulose, at the same volume fraction, without any processing, the efficiency of filtration may fall under 50 %, which leads to a considerable enrichment of fibres in the product and to an unacceptable quantity of recycled solids. It is then necessary to modify the characteristics of the fibres and/or the properties of the slurry to be filtered. This result can be obtained by various means, which are usually combined with each other

a) "refining" of cellulose, according to the technology of the paper industry, with conical or disc refiners. Refining makes the fibres more flexible.

b) use, blended with cellulose, of a small amount of polyethene fibrils, also called "synthetic wood pulp.

c) adding a flocculant stable at high pH (usually an anionic or non-ionic poly-acrylamide) to the slurry that artificially changes the granulometry of cement particles and increases their ability to be captured by the fibres during the filtration process.

d) some clay minerals like sepiolite or kaolin, seem to improve the situation a little further.

All these parameters properly combined allow to reach an efficiency of filtration of about 70%, which is far less than with asbestos and requires some additional modification in the process...

<u>**Basic problems arising from cellulose cement association:**</u>

Cellulose is the sole fibre basically compatible with Hatschek process. And yet, very deep differences exist between asbestos-cement and cellulose cement. Asbestos-cement has very stable properties over a considerable period of time (from a few days after manufacture to several decades), with a slow tendency to an increased brittleness.

On the contrary, cellulose cement is hardening (strength, stiffness) very slowly and is very flexible with amazingly high elongation at break at young age, especially when tested wet. Then it becomes elastic to rupture, brittle, with a strong tendency to cracking under the effect of thermo-hygrometric variations, within a few months or years (of course this behaviour is depending upon the kind and volume fraction of cellulose) ... This inconvenience is accepted for the limited internal use of this material, but never for external use of buildings.

Without asbestos, it is no longer compatible with the expected lifetime of cladding or roofing products. To prevent such a quick embrittlement, two ways have been developed in the period 1975/80:

a) limitation of cellulose content to a low value (about 4 ± 1 % of the total raw materials) and combination with one or several other fibres (called "reinforcing" fibres) to get the desired short- and long-term properties.

b) use of pure cellulose in a modified cement matrix.

We will successively analyse these two ways to asbestos substitution.

the two main routes for asbestos replacement

- use of a fibre blend

Many types of fibres have been studied to reinforce cement matrices. If we do not take into account fibres like carbon or aramid fibres which are very expensive and do not impart to cement products valuable characteristics (due to weak bonding between fibres and matrix), attention was focused on four main types of fibres that led to significant industrial development or production: alkali-resistant glass fibres, polypropylene and high modulus PVA or PAN. For use in the Hatschek process, candidate fibres must be stored, handled, dispersed in water, mixed with cement, in a way similar to asbestos or cellulose.

This provides a limit to the size of fibres: long, coarse and stiff fibres, like steel fibres, are not suitable.

The fibres must be chemically compatible with cement paste and give the end products the desired physical and mechanical properties (strength, toughness) with a good durability, at an acceptable price...

Polypropylene, under the form of staple fibres or chopped fibrillated fibre though having a good compatibility with cement, is hampered due to its low strength and stiffness as well as its weak bond with cement. It was rarely used in the Hatschek process, but produced new reinforced materials.

Alkali-resistant glass fibres have high strength and stiffness which make them theoretically suitable for reinforcement of cement. However, twenty-years of experience with "GRC" has shown the effects of aging on glass fibres, which result in brittle materials.

This difficulty could be overcome in "GRC" by adding polymers and pozzolanic materials. Such formulations are too expensive and not compatible with Hatschek process. Attempts to use dispersible glass fibres in combination with cellulose for making corrugated sheets or slates led to problems of brittleness and cracking and were abandoned.

The two fibres which led to the largest industrial success are high-modulus PVA and PAN. They have a good durability in cement paste (more than ten years-experience is now available) and high mechanical strength. PVA has probably the most adequate and stable bonding with cement...

PAN exhibits a slow embrittlement in cement on the long term. It is less strong but less expensive.

-Use of pure cellulose in a modified matrix

Autoclaving has been used for decades in asbestos-cement to make strong, thin flat sheets and pressure pipes. For flat sheets, the aim was to minimise moisture movement and carbonation shrinkage. Therefore, it seemed interesting to autoclave cellulose cement materials to prevent the embrittlement through continuous hydration and carbonation. Yet, autoclaved cellulose cement products, are less dense than asbestos-cement... and pressing is necessary for products submitted to harsh conditions, especially freeze-thaw cycles. Flat products in asbestos-cement are often pressed in stacks and the process applies to substitutes as well. For corrugated sheets, a single press with special filtering moulds, winding with the same cycle as the Hatschek machine, is necessary. Such presses have been recently developed since the process was exceptional for asbestos-cement. They represent huge investments (about 3 M US $) and cannot be installed on all existing lines...

Performances and durability of substitutes vs asbestos-cement:

All these characteristics are enhanced on unpressed products to an extent which makes pressing an absolute necessity for external application, with a possible exception for autoclaved cellulose cement flat or corrugated sheets used in dry climates.

Another typical weakness of substitutes is that they are much more sensitive to manufacturing conditions than asbestos-cement: the two major defects resulting from inadequate process control are a poor interlaminar bonding and an actual fibre content in the product different from the reference formulation.

Poor interlaminar bonding is particularly damageable to the durability since it may induce swelling and delamination under the effect of freeze-thaw cycles or surface cracks under thermal shocks.

Finally, one should mention that all organic substitutes to asbestos (cellulose, polyethylene, polypropylene, PVA, PAN) are highly combustible...
It seems that raw material costs are an increasing part of the total direct costs (from 62 % for asbestos-cement to 75 % for air cured cellulose/PVA cement) and it is quite unlikely that this situation would improve in a near future.
The amount of additional investment for one production line is about 4 M US $ for the air curing process and 6 M US $ for the autoclaving process, both including a single press.

There is no evidence that the market would accept such an increase in the price of roofing fibre cement products, the price gap between asbestos-cement and other substitutes like galvanised steel being currently of around 10 to 15 %.
Consequently, the major risk associated with substitution is that the market share for fibre cement products is increasingly decreasing, as has already been the case in some countries.

- NEW TECHNOLOGIES

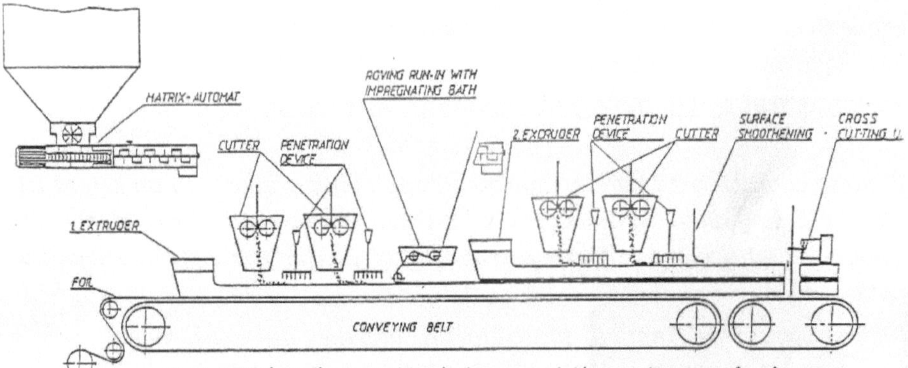

Figure 5 Schéma d'une machine à plaques ondulées en ciment renforcé verre (GRC). Le système d'ondulation n'est pas montré.

05-002 Scheme of a corrugated sheet machine for glass reenforced cement

processes derived from manufacture of "glass reinforced cement"

Manufacture of "glass reinforced cement" (or concrete) was set up at the Building Research Establishment, in the UK, in the late 60's, from techniques used to make glass reinforced plastics (13). Its development was made possible with the use of the so-called "alkali-resistant" glass fibre, containing about 15 % Zr02. The main processes are simultaneous spraying of fibres and cement, or moulding of a premix.

a) *mechanised production of flat sheets.*

These products were developed in the U.K and in Japan in the late 70's. The process consists in laying or spraying chopped fibres and cement paste on a continuous conveyor with, if necessary, a dewatering system.

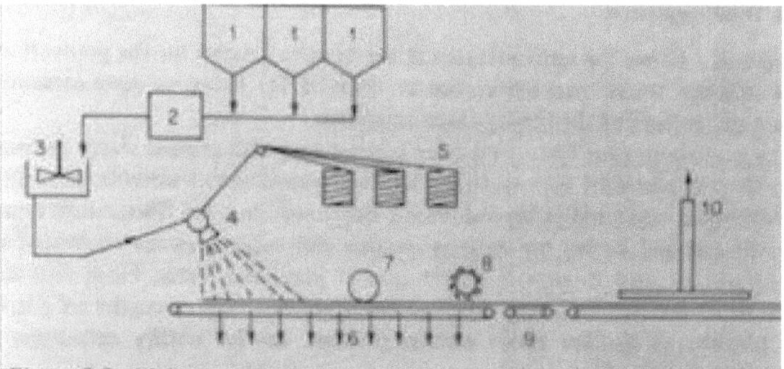

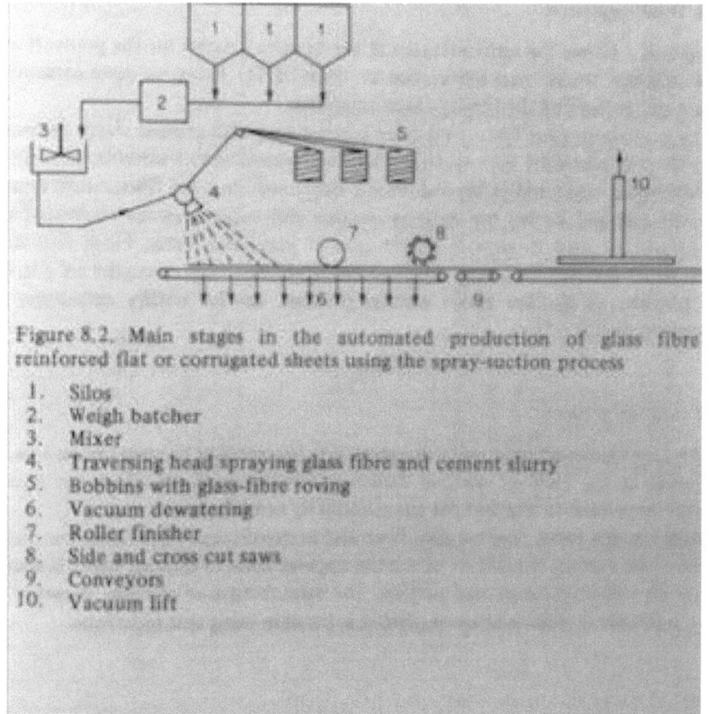

Figure 8.2. Main stages in the automated production of glass fibre reinforced flat or corrugated sheets using the spray-suction process

1. Silos
2. Weigh batcher
3. Mixer
4. Traversing head spraying glass fibre and cement slurry
5. Bobbins with glass-fibre roving
6. Vacuum dewatering
7. Roller finisher
8. Side and cross cut saws
9. Conveyors
10. Vacuum lift

05-003 Spray (GRC). On both schemes corrugating devices are not shown

b)mechanised production of corrugated sheets

A more sophisticated application of this process was recently developed in South Germany for the manufacture of corrugated sheets (14) and is shown on figure 5: the product consists in two layers of 3.3 mm of cement mortar and chopped fibres laid on a plastic film supported by an endless conveyor. Continuous glass roving is incorporated between the two layers at places which will become the valleys of the corrugated sheets. After curing, the sheets are turned over and the inferior layer which is becoming the apparent face can be through coloured. Due to its contact with the plastic film, aspect surface is very smooth and glossy. The products have fairly good mechanical properties (flexural strength, impact strength) at young

age. However, there is a doubt on the durability of these characteristics. After storage in stacks on stockyards some fine transverse cracks were noticed, as well as surface micro-cracking. This phenomenon might result from the difficulty to get a homogeneous distribution of the reinforcement at the edges or on the surface of the sheets.
The sheets are rather thick and very dense (1.95 g/cm^3) so that the weight of the roof is about 30 % higher than with asbestos-cement roofing sheets.
Another problem arises from the fact that the associated moulded pieces (ridges, eaves...) cannot be manufactured with the same process. The cost and higher consumption of raw materials result in a considerable increase in cost compared to asbestos-cement corrugated sheets.

It seems therefore doubtful whether this technique will significantly extend in the future at the expense of Hatschek process.

Extrusion The required technics are quite complicated and could hardly be developed in a book mainly dedicated to a historic facet. Just for the pleasure of the non-specialist reader let us have a look to the following diagrams of extrusion devices, without any more comment.

Extrusion of asbestos-cement was developed in the early 60's, initially by Johns-Manville Corporation from the technology used in the clay industry (15) to make plain or hollow profiles for windows and partitions

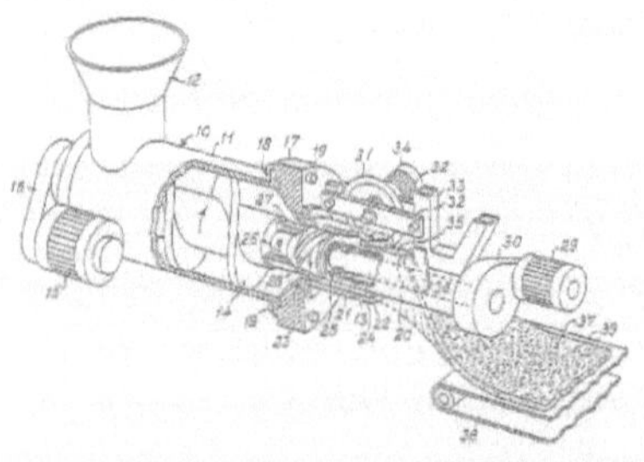
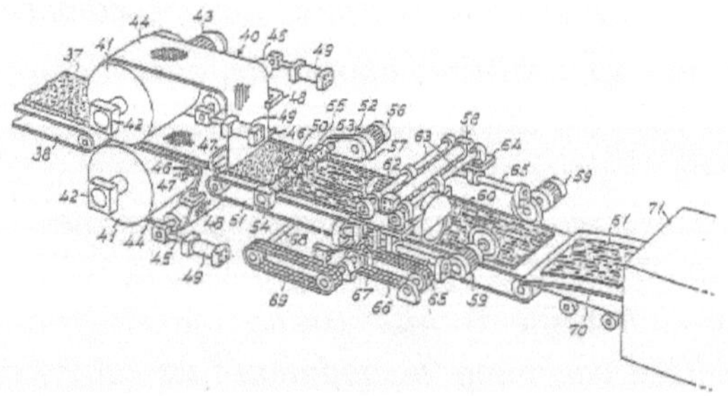

05-005 & 05-006 Examples of Systems conceived for extrusion process.

Due to their high prices, these products gained a limited industrial success and were almost abandoned, until the process was recently adopted in Japan to manufacture a wide range of profiled products with various formulations, most often containing asbestos. Another interesting development took place in Denmark for manufacturing non asbestos slates..."

Process using polypropylene fibrillated films:

Reinforcement of concrete with discontinuous elements of polypropylene is not recent, but the idea of using a continuous reinforcement of cement under the form of polypropylene fibrillated films was originally patented in 1976 by the University of Surrey in the UK described by Hannant ...

...Other processes, using woven polypropylene reinforcement (20) or a combination of polypropylene and glass fibres (21) were later proposed...
From the laboratory work at Surrey, a first attempt to develop an industrial process for manufacturing corrugated sheets, was carried out in the Netherlands and failed. Several other attempts were undertaken in Italy in the early 80's and are in progress...
...One of the major difficulties of this new technology is to get a homogeneous distribution of the reinforcement throughout the thickness of the material, as well as a smooth and defect free surface aspect of the corrugated sheets...
Recycling green or hardened waste, either in separating the polypropylene network from cement or in grinding them together, still raises difficult problems.
Finally, this process does not allow the production of moulded parts.
At the present stage of its development, the production efficiency of this new technology does not exceed 50 % of that of a modern line of asbestos-cement.

- CONCLUSION

This review of substitutes to asbestos-cement building products leads to the following points:

a) *Hatschek process can be converted to alternative fibres, id est cellulose fibres (pure or blended with polymer fibres). But this transformation requires costly additional equipment, expensive raw materials, and skilled labour for running the process. The quality of finished products is very sensitive to manufacturing conditions.*

b) *Flat sheets and roofing slates are products for which asbestos replacement raises less problems. About 10 years of field experience show that substitutes can meet the market requirements. For slates in particular, the high level of price of natural slates in northern Europe allowed substitutes to take a significant market share.*

c) *at the opposite, asbestos replacement would result in a dramatic challenge for corrugated or profiled sheets, which represent by far the most important production of asbestos-cement in the world: huge investment, increase in cost up to 75 %, limited knowledge of the performances of substitutes. In some cases, as long sheets with deep profiles which are currently used in many countries of Latin America, there is no existing substitute to asbestos-cement.*

Asbestos-cement still remains the roofing material having the best performances at the lowest cost.

d)*New processes that have been developed in the past ten years do not seem to be able to take over Hatschek process in a foreseeable future. Each of them is limited to particular materials suited to market niches that would accept a high level of price."*

At the time of this big international meeting, in spite of the fact that many asbestos-cement machines were already at standstill, a great part of asbestos imported in France was used in asbestos cement products, by Eternit in its plants of Thiant, Saint-Grégoire, Albi and Paray-le Monial, and by Everite with its factories of Dammarie-les-Lys, Descartes and Andancette.

But evolution was going fast. In 1995, seven European countries had already banned the use of asbestos as well as the sales of products containing asbestos. *(Germany, Netherlands, Switzerland, Denmark, Sweden, Norway and Italy).* In these countries, Eternit factories had replaced asbestos by cellulose. But in France, asbestos-cement manufacturing was still going on, though reduced if compared with previous years. Some experts explain that at the time France had become the biggest European asbestos-cement producer. However, due to the increasing number of alarms, the Labour Department and the Health Department ordered a report to INSERM *(National Institute for Health and Medical Research).* The report was published on July 2nd 1996. The main parts of the conclusion, in form of a 69 pages synthesis, were immediately made public. See "the main dates of asbestos".

In the following days, the French Government, based on the report, issued a decree which banned the use of asbestos from January 1st 1997. Industrial companies had to accept the injunction, which came as no real surprise to them and for which they were partially prepared.

Associations that defended victims were enraged and were supported by the press .Newspapers were happy "to talk about. Trials multiplicated, lawyers displayed their eloquence and made profits... Victims were add up those already dead and those to come, while finance departments were making provisions for the future, sometimes at the cost of investments. The Metro *(Paris Underground*) did not escape the crisis. At the time of the all-powerful asbestos, sheets and pipes had been displayed in various uses, in ventilation, in recovering places to protect or to hide, etc. Between 1975 and 1985 RATP[90] led a big campaign in order to remove asbestos from its tunnels and so numerous corridors, but it did not actually calm down the fears of the maintenance department[91]workers.

At the end of this chapter, one can find a chart from "International Ban Asbestos Secretariat" compiled by Laurie Kazan-Allen, about countries having definitely banned asbestos.

---

[90] RATP : Régie Autonome des Transports Parisiens, public entity in charge of public transports in Paris and around.

[91] Information from the French newspaper Liberation, 1995 10 17.

In 1997, Saint Gobain gave birth to a new subsidiary, NovaTech *(for New Technology)* which signed the death of the old Everite, and counted three factories:
-Descartes, recently transformed with the arrival of the Wellcrete technology.
-Dunkerque, near the Channel, Hauts de France, with the same type of machine as in Descartes, and another one, home conceived, due to produce glass reinforced slates by extrusion.
FBK *(Faserbetonwerk*[92] *Kalbermoor, in Kalbermoor, Bavaria, Germany),* purchased from Heidelberger Zement who invented the technology.
Unfortunately, for the young NovaTech, born from the ashes of the old Everite, the technical choices were probably not the best ones. Continuing partnership with FBK and HZ would perhaps have offered more favourable results. However, let us remember that afterwards judgements always are easier than the right choice in the moment. Big investments were lost and soon the company disappeared from the French industrial panel, giving space to Eternit, his eternal competitor. We will see in chapter VII "Wind of Change" that thanks to the experience acquired in neighbouring countries, Eternit made another choice.
In other countries, choices were diversified, influenced by environmental exigencies from people, states behaviour, availability of local raw materials. In some cases due to a complete lack of interest on the part of the government or even simply because of its own citizens. More details will appear in chapter VIII "Country Overview".

"...The World Health Organisation has included asbestos in drinking water from asbestos cement pipes in its 1993 edition of the Guidelines for Drinking Water Quality. The guidelines state "Although well studied, there has been little convincing evidence of the carcinogenicity of ingested asbestos in epidemiological studies of populations with drinking water supplies containing high concentrations of asbestos. Moreover, in extensive studies in laboratory species, asbestos has not consistently increased the incidence of tumours of the gastrointestinal tract. Therefore, there is no consistent evidence that ingested asbestos is harmful to health and it was concluded that there is no need to establish a health-related reference value for asbestos in drinking water". Asbestos cement pipes have been widely used for drinking water distribution and there are many kilometres to be found all over the world. Although few countries

[92] Faserbetonwerk : fibre reinforced concrete.

still install asbestos cement pipe, primarily because of issues with handling, there appears to be no concern for health of consumers receiving the water and no programmes to specifically replace asbestos cement pipe for this reason."

Copies of this report are available as an Acrobat pdf to download under the 'Post 2000 Reports' heading of the **Research Page on the DWI website.** In 2003, a memorial was unveiled in Thiant, Hauts de France, birthplace of the French Eternit. Later on, several towns proceeded with similar decision

05-008 Memorial to the victims in Thiant-North of France- were production of asbestos cement began early 19th century.

Here is a copy of International Asbestos Ban Secretariat, dated October 2016. A revised updated copy can be obtained on: Worldwide asbestos ban.

If you want to have less pessimistic ideas about a disappeared product and work with the grief of the victims of the Satanized fibre, let's visit an often-unknown aspect of fibre-cement: its close connection to art and architecture

International Ban Asbestos Secretariat

Current Asbestos Bans and Restrictions

compiled by Laurie Kazan-Allen

(Revised October 16, 2016)

National Asbestos Bans:[1]

Algeria	Denmark	Ireland	Mozambique	Seychelles[3]
Argentina	Egypt	Israel	Netherlands	Slovakia*
Australia	Estonia	Italy	New Caledonia	Slovenia
Austria	Finland	Japan	New Zealand	South Africa
Bahrain	France	Jordan[2]	Norway	Spain
Belgium	Gabon	Korea (South)	Oman	Sweden
Brunei	Germany	Kuwait	Poland	Switzerland
Bulgaria	Gibraltar	Latvia	Portugal*	Turkey
Chile	Greece*	Lithuania*	Qatar	United Kingdom
Croatia	Honduras	Luxembourg	Romania	Uruguay
Cyprus*	Hungary*	Malta*	Saudi Arabia	
Czech Republic*	Iceland	Mauritius	Serbia	

Notes. Singapore and Taiwan were removed from the ban list (Oct 2010) as a result of information received.

Mongolia was removed from the ban list (August 2012) as a result of information received detailing the cancellation on June 8, 2011 of the Mongolian Government's Resolution No. 192 banning asbestos which was issued on July 14, 2010.

---

[1] Exemptions for minor uses are permitted in some countries listed; however, all countries listed must have banned the use of <u>all types</u> of asbestos. Additionally, we seek to ensure that all general use of asbestos, i.e. in construction, insulation, textiles, etc., has been expressly prohibited. The exemptions usually encountered are for specialist seals and gaskets; in a few countries there is an interim period where asbestos brake pads are permitted.

[2] An immediate ban on amosite and crocidolite was imposed on August 16, 2005; a grace period of one year was allowed for the phasing out of the use of tremolite, chrysotile, anthophyllite and actinolite in friction products, brake linings and clutch pads. After August 16, 2006, all forms of asbestos were to be banned for all uses.

[3] In 2012 it was reported that due to lack of enforcement, asbestos-containing products were still being imported and used in the Seychelles.

* January 1, 2005 was the deadline for prohibiting the new use of chrysotile, other forms of asbestos having been banned previously, in all 25 Member States of the European Union; compliance with this directive has not been verified in countries with an asterisk (*). As of May 2009 there are 27 Member States, with Romania and Bulgaria joining the EU in 2007.

05-008 Memorial to the victims in Thiaant- North of france-where production of asbestos-cement began ealy 20th century

VI Fibre-cements, Art and Architecture

"Gentlemen, managers of asbestos-cement, you have plenty to do ahead, be delighted and delight us: you still can open lots of doors and lots of windows to the fancy of architects and of all those who, thanks to their talent, participate with them to the charm of architecture, which would not deserve its name, would it not be the charm of our life".

Guillaume Gillet[93]

From the moment asbestos-cement was invented, artists discovered what this new paste could offer them. Cheap, malleable when fresh, hardening quite fast and very tough once hardened, it could offer a lot in sculpture and all kind of moulding.
Guillaume Gillet wrote in his preface to *"Le Fibrociment dans l'Art 1903-1973"*, published by "Le Fibrociment Elo":
"Two French engineers, the Lanhoffer brothers, held an important part in the genesis of asbestos-cement industry. Soon after the opening of the Austrian factory in Vöcklabruk, they purchased Hatcheck licences and created "Société du Fibrociment" in Poissy" -some 40 km west of Paris. *"Since 1903, the manufacturing of fibre-cement remains based on the same process, but numerous improvements were made, permitting to obtain new and better products*

[93] Guillaume Gillet (1912-1987) Premier Grand Prix de Rome, Chief Architect for French Public Buildings and Palaces. Great French architect of the 20th century who, among others, in cooperation with Marc Hebrard, built the famous church of Royan, Atlantic cost, France, in 1947and the Palais des Congrés, Porte Maillot, Paris.

Société du Fibrociment marketed all types of asbestos-cement products but specialized in mouldings and sophisticated boards, both for inner and outer architecture, under the make: Elo. Glasal[94] *certainly was the Company's flagship.*
No surprise then that artists and designers made a wide use of the product in every form"[95].

According to Guy Delas, great historic expert in the matter, a Swiss firm who owned an industrial site in Poissy, created there a flourishing company specialized in asbestos-cement products. Quote: *"On September 7th 1903, the Compagnie Continentale d'Electricité registered Fibrociment de Poisssy as a trademark with the Court of Versailles, under number 641. It included artistic models and patents in the name of Fibrociment Gallia Poissy"*[96]

06-001. Fibrociment Gallia.
Colección Olivier Delas.

Driven by Oscar Lanhoffer, his visionary manager, the new company knew a fast and impressive development, especially in the field of ornament boards. The company's name, "Fibrociment" was so famous that it soon became the generic name of asbestos-cement, somewhat as the name of Kleenex became the popular name of nonwoven handkerchiefs.
Olivier Delas, goes on: *"Before the 1914s, a new activity began to grow apart of utilitarian manufacturing. It turned towards inner adorning products. Asbestos-cement*

[94] Glasal: Very famous fibre-cement product which will be widely detailed in the following pages.
[95] Adapted from the French by Jacques Roulland.
[96] Most details about « Fibrociment de Poissy » are quoted from « L'établissement Industriel du Fibrociment de Poissy » by Olivier Delas.

*boards were then decorated using a technic known as "patina" or "glaze", which later on became "Glasal". With such a mineral enamelled layer, adorning panels met a real success under the name of the trademark "ELO" built from the initials of the two brothers (**E**dmond, **L**anhoffer, **O**scar) patented in 1920. The product soon knew a great development in terms of protective and decorative boards on the front of numerous shops". great development in terms of protective and decorative boards on the front of numerous shops".*

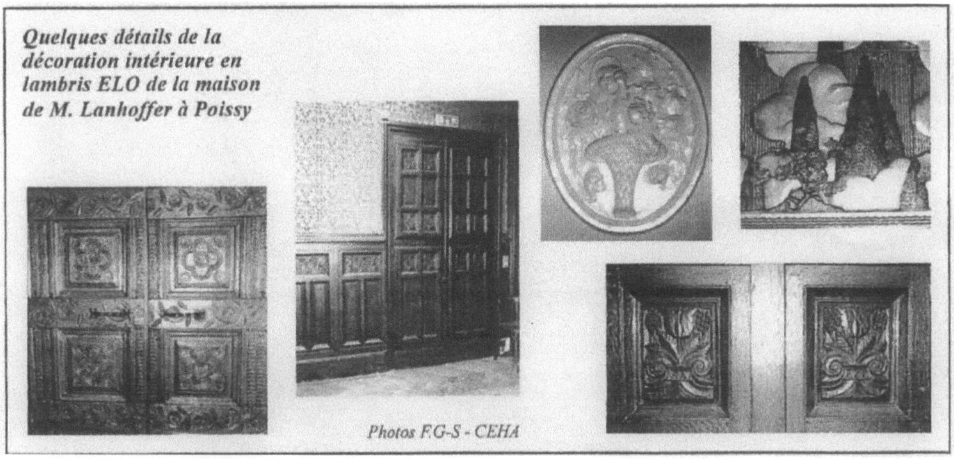

06-002 details of Lanhoffer house. Olivier Delas

Still referring to Olivier Delas historic research, let us note that the factory was using a high percentage of female workforce, "*but the company was caring about its members. They structured systems such as children watching. Many workers belonged to Italian families. When wife and husband were both working in "Fibrociment", their children were taken care of in the company's day-nursery on site. A worker's estate was built on a plot next to the factory. It was known as the "Italian City". Status included a right for the workers to purchase their dwelling... As soon as 1913, Mister Lanhoffer also organized a food cooperative, where workers could purchase most needed alimentary goods. A restaurant was even set in premises let by the company, it supplied meals to members...*

Decorative panels knew a great success with a high manufacturing development. An advertisement of the time stated: *"Thanks to its perfect wood imitation both by concept and fineness, ELO claddings totally imitate wood, with the advantage of being fireproof and of keeping their initial shape. Maintenance can be made just like with home furniture, just using CIRELO[97]".*

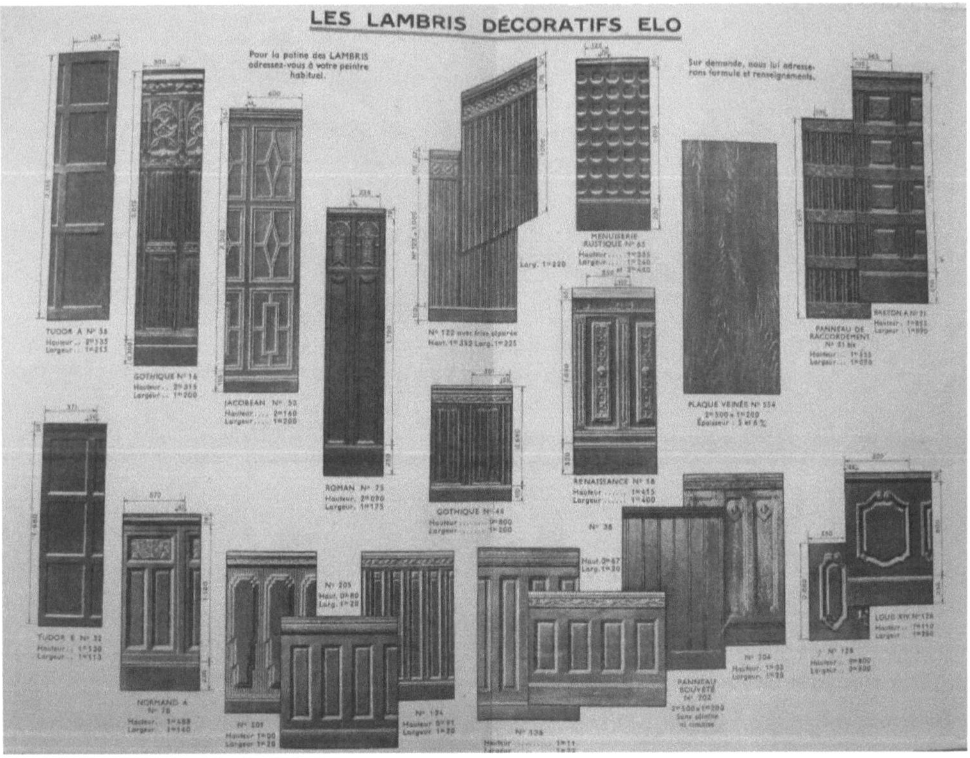

06-003 Decorative ELO panels. Olivier Delas

Until the 1928, expansion was indeed limitless, but when the 1929 crisis arrived, things began to change. As we could see previously, the already ambitious Eternit had purchased Fibrociment de Poissy and Elo as soon as 1930. At the same time Eternit also purchased Ouralith near Toulouse -south of France- and closed it soon

[97] CIRELO sounds more or less like ELOWAX!

due to the advanced state of dilapidation of the factory. In the contrary, Fibrociment de Poissy and Elo knew a prosperous period, including a move to the nearby Triel, where Eternit was already established, with corresponding investments.

In 1936, the eleventh *Salon des Arts Ménagers* (Ideal Home Show) included a large exhibition dedicated to ELO's asbestos-cement decorative panels. All the staff and concessionaires of the company made the most of the opportunity to afford a family snapshot on the steps of "Le Grand Palais"[98]. Before leaving this likeable company, let us note that in 1941, when France was occupied by the German army, a group of industrialists from Poissy created a training centre for young men. Some eighty pupils were welcomed in the boys' school where they received technical schooling and in one of "*Fibociment's*" workshops where they received practical training. In addition, the system helped many pupils to avoid enrolment in STO[99].

In March 1942, the allied bombed the Ford factory in Poissy, some lost bombs fell on the "Italian city" which was partly destroyed. On request of "Fibrociment" management, German militaries who stayed in nearby undamaged buildings accepted to leave, which allowed some victims of the bombing to obtain new accommodation. Many French people still have the relief map of France in mind that hung on walls of their classroom, such as the one appearing here which helped them learn names of rivers and mountains of their country. The map was due to "Fibrociment de Poissy"

06-004 Relief map of France

[98] Le Grand Palais: Famous Paris exhibition show building located near Champs-Elysées and the River Seine.

[99] STO: *Service du Travail Obligatoire:* Compulsory Work Service, created by the French and used by the Germans during WWII to compel thousands of French workers to travel to Germany as forced labour for the German war effort.

;

We must also stipulate that Olivier Delas, our generous informer, came to be deeply involved in the history of “Fibrociment de Poissy” by his 25 years fight against authorities, to obtain that the illness that killed his father was considered as a professional disease.

Around 1919, Georgette Agutte, French sculptor and painter, produced “**La Robe Bleue** ”,an oil paint on a fibre-cement sheet, now exposed in Grenoble Art Museum. I planned to incorporate it in the cover of the present book. Unfortunately, the museum was too avid.

06-005 Portrait by Georgette Agutte painted on fibre-cement sheet. Courtesy of La Maison Sembat Agutte; Bonnières sur Seine 78270

I was luckier with La Maison Agutte Sembat de Bonnières sur Seine, Yvelines, France. There, the curator was very happy to supply me with a picture of one of her works, which though not so famous, was also painted on a fibre-cement sheet.

At the beginning of the 20th century, Spanish Catalonia was also very energetic in arts as well as in industries. The Roviralta family, mother of the future Uralita, was intimately linked to the artistic Italian Lena family lately arrived.

06-007 Tres ejemplos de estatuas del escultor LENA, realizadas en los talleres de Uralita en Cerdanyola del Valle. Cortesia del Museo de Arte de Cerdanyola.

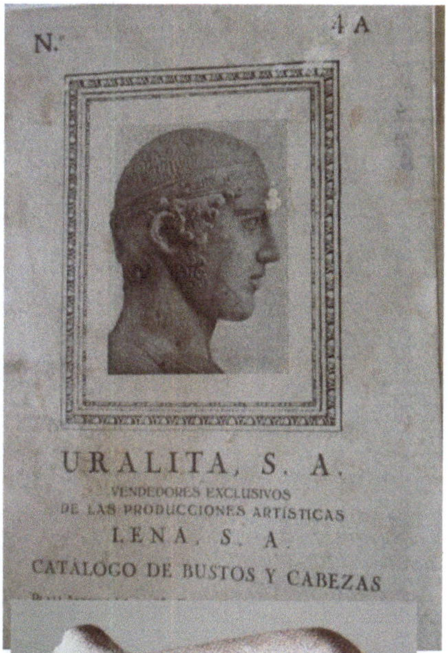

06-006 Fibrecement tortoise tile conceived by Antonio Gaudi

06-007 Lena statue for a Barcelona park.

Oscar Lena was searching a place where to be at ease with the tall statues he was preparing to adorn squares and gardens of Barcelona. In the new factory of his friends, in Cerdanyola del Valle, he found both a workshop and a new and cheap material duly fitted to his needs of outer protection for his plaster productions.

Not only did he brilliantly made the most of his discovery, but his activity gave ideas to the managers of the company who soon developed various decorative products more or less like Fibrociment de Poissy and Elo where doing in France. Sculptures, inside and outside decorative panels as well as all kind of adornments flourished till 1936, when the Spanish civil war began whose sinister stories remain in everyone's memory. Every year, Uralita was publishing catalogues praising its decorative novelties. Ancient works of art from Roma and Athens were duplicated and sold in countless copies, as can be seen on the enclosed picture number 06-006.

The famous Catalan architect Antoni Gaudi[100] was also interested in asbestos-cement possibilities. He conceived an original tile, named "tortoise tile". I could not find one, if it was ever used in architectural works.

In 1929, Uralita had been actively engaged in the Barcelona International Exposition. Among others, they displayed a prefabricated pavilion conceived with asbestos-cement sheets, which actually called vis-

[100] Antoni Gaudi: 1852-1926, famous architect father of the Sagrada Familia Church and numerous typical houses in Barcelona.

itors attention. Lots of statues embellishing Barcelona's parks, enhanced on the occasion of the Exposition, originate from artists who worked with asbestos-cement paste coming from Uralita's factory. Some of them were rehabilitated and relocated on the opportunity of the 1992 Olympic Games held in the town.

06-008 Uralita fibre-cement pavilion .

Uralita was more and more engaged in street furniture which could be met everywhere in Spain during many decades all along the 20th century. Such was the case, for instance, with public pillar-box obviously inspired by the famous British red pillar-box and also with ONCE's[101] ticket booth.

06-009 Post office box

06-009 Once Fibre-cement ticket office (Uralita 80 años).

[101] ONCE: Oficina Nacional de los Ciegos Españoles: National Association of Spanish Blind People.

Whatsoever, Uralita did not lose interest in inner decoration as testified by the following advertisement.

06-011 Fibrecement decorative panels by Uralita

Josep Sert, another famous Catalan architect, tells us[102]:

"With my friend John Miro, I often had long talks about mural painting and its relation to architecture. I have had the privilege of collaborating with him on several occasions, and Miro who talks less but thinks more than painters has found a new approach to mural painting that gives up the idea of the continuous mural, replacing it by a more logical treatment of a wall by focal points.

102 Extracted from "In their own words, correspondence 1936-1980" by Patricia Junseca Vecchierini (Editor).

*He started this new line back in 1935 in Barcelona. We had formed in 1931 a group of young architects that were interested in integrating their work with that of painters and sculptors that had also broken with the past. This group set up a small exhibit in the annual "*Salo dels artistes decoradors*" and we asked John Miro to work with us. We collaborated from the start. On the small wall surface available Miro outlined a free shape related to the wall but covering only part of it. For this he took an asbestos sheet and had the shape cut out as outlined, he then made a very lively painting of mural quality that left parts of the asbestos panel apparent. This painting was set in the plaster of the wall that Miro has painted in a colour of his choice.*
He told me years later that this started him on a new road towards mural painting".

06-012 Fibrecement Coffered ceiling at Cerdanyola Town

During my visit to the Art Museum of Cerdanyola in 2016, the curator (I want to greet his collaboration), helped me to the town hall. He asked to open the counsel room in order to let me look at the wonderful coffered ceiling. It had also been produced by Uralita in the first half of the 20th century.. When she heard that the ceiling was indeed made of fibrecement, even though she thought it was wood, the young clerk who had brought the key was deeply upset. She started running through offices, telling the news to everyone around. I ignore if employees urged the council to replace the ceiling.

In the 1930, architects also used fibre-cement pipes as decorative and sustaining columns. In Angouleme, France, a doctor of medicine built his house with two fibre-cement pipes supporting a front canopy. Both pipes are 3 metres (118in) long and 0.30 metres (11,8in) in diameter. In 2018, the columns do not show any sign of fatigue. According to probabilities they had been filed with sand or gravel before being put in place. I did not dare to ask the owner to make a hole in anyone to confirm...

06-012bis Fibrecement pipes used as colons on a 1930 villa in Angoulême France

-

In 1939, at the New-York City World'sFair Johns Manville had a very wide space at his disposal. There they built an impressive building.

08-013 New-York World Fair 1939. Various origins

At the end of WWII, Pablo Picasso could not find any convenient canvas to let his talent run free. Guillaume. Gillet, already mentioned, reports a conversation between Pablo Picasso and Dor de la Souchère[103], dated September 8th 1946

"... I always wished to paint on very big areas, I was never offered any...

-Areas? You want areas? I can give you areas."

He gave a black look on me.

"But where can you get them?

- *In Antibes Museum"* - Antibes, Mediterranean shore, near Nice, South of France
- *The following day, we were running (...) Antibes' shops searching material (...) We bought every fibre-cement sheet".*

On September 17th, they actually purchased six Eternit flat sheets, each 120 x 250 cm (47in x 98in) at, 'La Société Cannoise de Matériaux', for an amount of FRF 2 256[104] as testified by the invoice scanned by Kimberley Muir in 'Picasso Express'".

This is how Picasso Museum in Antibes, nowadays displays several works of the Master painted on fibre-cement. Let us hope that developers of the zero risk will not convince one of our next Ministry of Health to have them destructed. Among works of art painted on the material, let us note: "Ulysses and the sirens", "La Joie de Vivre" and "Still Life with Bottle, Sole and Ewer". If you are lucky enough to travel by Antibes, do not hesitate! Pay a visit to the Museum, you will not regret it!

In 1952, Miro[105] was in charge of producing a monumental work planned to embellish the restaurant in Paris "Ideal Home Exhibition at Le Grand Palais". He also used fibre-cement sheets as a support. Another one by Robert Pansart was in the hall of Champs-Elysée Cinema Marignan. Unfortunately, this smart wall was destroyed on the occasion of a rehabilitation of the hall.

---

[103] Romuald Dor de la Souchère (1888—1977) teacher in French, Latin and Greek; archaeologist and friend of Picasso. He offered the artist a workshop in Antibes castle, where to develop his works. The castle later turned into "Picasso's Museum".

[104] FRF 2 256: maybe something like 6.5 USD from the time

[105] Miro: (1893-1993) Very famous Spanish Painter, one of the greatest in surrealistic art.

06-015 Painting on fibre cement sheets by Miro for the “”Ideal Home Exhibition at Le Grand Palais in Paris in 1952

06-016 Painting on fibre cement sheets by Fernand Leger for the restaurant of Gare de l’Est in Paris

06-017 Fibre cement wall by Robert Pansart in the hall of Le Marignan Movie Theatre , Champs-Elysées. Paris around 1950. Unfortunately, this smart wall was destroyed on the occasion of a rehabilitation of the hall.

Unsatisfied with roofs and walls, fibre-cement makers undertook with excitement to cover floors, mostly in water rooms and corridors, both in public buildings and private homes. At Everite, Bassens factory, near Bordeaux, France, which at the time still ran its own R&D, a decorative product named "Evergranit" was conceived. The support was a flat sheet of fibre-cement. Small pellets were sprayed on the forming sheet, by a paddle wheel feeder, right before the arrival of the last mono-layer to the forming-roll. The pellets were made of asbestos, white cement, mineral pigments in different colours and water, they would form a kind of roughcast on the surface of the sheet roughcast on the surface of the sheet roughcast on the surface of the sheet.

Passing under the weight and pressure of the forming-roll, pallets were flattened and crushed. Colours would be spread at random. One can imagine that the pallets launched by the laboratory engineers, reminded by our model maker, were not uninvolved with research regarding the design of the product.

The flat sheet was then sanded in order to obtain the required soft surface.

The new product knew a large success, overall in the south-west of France. It was mostly used for shops front windows at a time when wide glass panels were not yet ubiquitous. One must admit that it was very resistant in the long term as proved by the next picture where it was used as a kitchen floor. Laid around 1960, it still looked like new, though somewhat old fashioned when I took the picture in 2017.

In 1954, in Biot, South-East of France, a group of artists and architects, organised an exhibition in an olive grove. They named it Architecture, Shapes and Colours. Reminding this utopian exhibition, Biot's Art Museum displayed two paintings produced by the already mentioned Fernand Leger 62 years earlier.

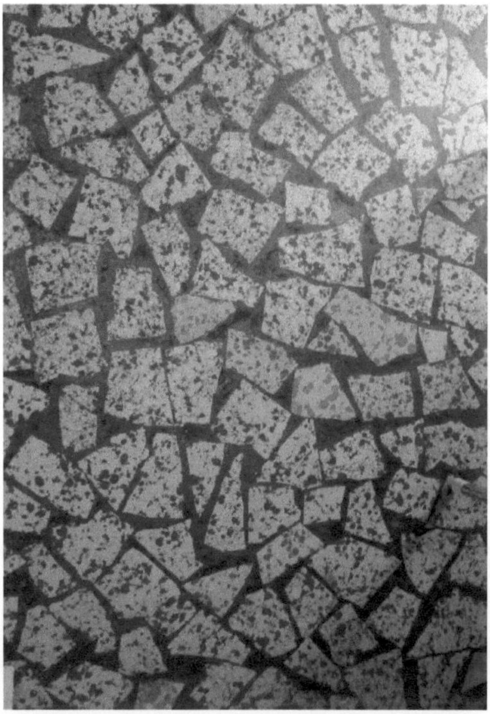

06-018 Evergranit 60 years after laying in the of worker at Everite, Bassens factory

Unfortunately, I had no opportunity to see this exhibition and I could not retrieve the corresponding pictures.

For his part, Eternit developed garden and inner furniture in several European countries, mainly for use in kitchens and bath-rooms.

Still at the same time, whilst fibre-cement was in vogue, artistic creations were cheerfully going on. For instance, Willy Guhl, famous Swiss designer (1915-2004), discovered the skills of fibre-cement. He made the most of the flexibility when fresh and with the resistance when dry, to design various decorative garden items. Lots of them are very famous, though people often ignore how they were made and with

which material. His best-known creation is certainly the "Loop Chair", but his "Handkerchief Flowerpot" and many others also deserve our attention. A visit on the Internet might offer you some surprise as to current prices of the mid-20th century fibre-cement works.

-Years later, after the arrival of "new technology" without asbestos, Willy Guhl designed a new" Loop Chair" with the same success.

Any visit to "für Gestaltung" Museum in Zurich, Swiss, offers a clear view of the numerous tal

06-019 Kitchen by Eternit

of this artist and of the possible works which can be obtain with fibre-cement material.

06-020/06-021 Willy Gulh loop chairs

In 1958, Guillaume Gillet, already mentioned famous architect, conceived the French Pavilion in Brussels International Fair. He "*wanted to suggest a building that would be the equivalent of "the Machines Gallery" in 1889".* He covered the inner dividing walls with rough fibre-cement sheets, modelled from his own drawings, under advice of the best acoustician." *It concerned a 200 seats hall, fitted with stereophony, and we were satisfied"* did he write in 1973. He thought to do the same in "the Palais des Congrés at Porte Maillot" in Paris when he was in charge of the auditorium *"with a twenty times wider hall..."* But that was another story, and it seems that the previous technic was eventually not taken on.

In 1964, Eternit Switzerland again had its own pavilion at the National Fair in Zurich.

In our 21st century, Eternit becomes aware of the role played by fibre-cement in architecture. The above website and its QR code offer a good sight of the various involved works.

Glasal.

We already said a few words about Glasal in previous pages. It is about time to tell you more about, due to the important part played by the product in the building industry during the decades following WWII.

Invented in Switzerland between the two world wars, Glasal was improved by Eternit Belgium in its Kapelle-on-den-Boss factory where it was developed in 1954. In France, Fibrociment de Poissy purchased the licence in 1957. Immediately they built a new factory in Triel sur Seine, west of Paris and near their original factory. The idea was intensive production, in order to meet the huge demand coming from architects in charge of the numerous, and urgent, large residential suburban buildings of the time.

Let us refer to the dissertation of an architecture student[106] and let us see what she tells us about Glasal: *"air-steamed and sanded flat sheets (fig.1) which receive a first colouration according to a 'secret formula' (fig.2). A short time in the oven makes sheets easy to handle and make them ready for their varnishing. (fig.3). After a second time in the oven for complete drying (fig.4), they are soaked into chemical baths*

106 Françoise Lampaert, « Glasal: colored, multipurpose and sustainable material. Year 2009-2010, National Architecture High School, Lille, France.

06-021 Glasal manufacturing steps.
Essay: « Le Glasal ». Françoise Lampaert 2009-2010.

which definitely secures the colour (fig.5). The last operations --washing, moving around, rinsing, drying-- (fig.6) give the sheets their smooth and steady look...[107]

The new product counts with numerous qualities, I fear to forget some of them: Apart of low prices and easy use, one of the first distinctive features that seduced architects, was the richness of the colour chart. In 1961, it offered 15 shades, in 1969 it offered up to 25. Let us remember that we were then in a time of accelerated building of large housing blocks in order to face-up to the arrival of people to big towns and their suburbs. Everyone began to be bored with the "cage look" of large suburban buildings. Anything that could help was welcome. According to Eternit, Glasal was the most suitable product to answer the new common wish.

"While years were passing, the product revealed its multiple qualities: ability to support temperature change, especially in freezing times. Non-combustible, mechanic strength, endurance to dirt and chemical attacks. Such qualities allowed glasal to penetrate kitchens, bath-rooms, surgeries. Work-surface indifferent to any aggressive stuff were a wonder! It was water-repellent, insulating (λ=0.3W/m°K), *soft and rot-proof. In spite of their stiffness, Glasal panels were easy to handle with usual tools such as handsaw or drill.*

On request of architects they could easily be obtained in different sizes and thickness for specific works".

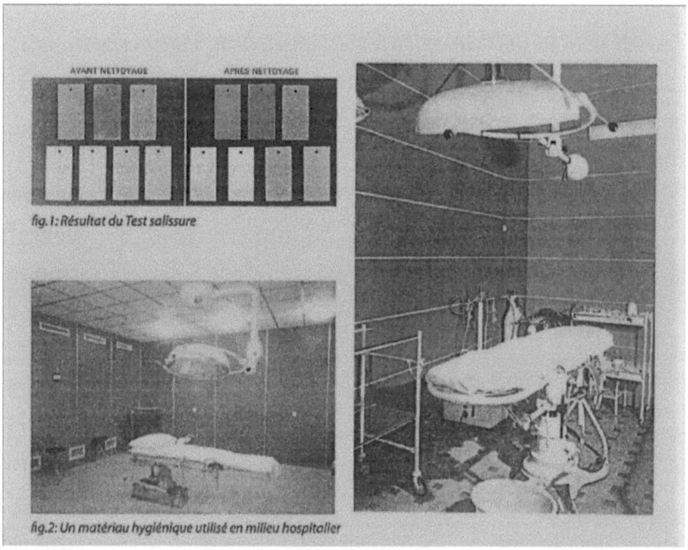

06-022 Surgery rooms

---

[107] Different sizes are available: Length: metres 2,520- 3,070- 3,200; inches 99,2126-120,86- 125,984/ Width: metres:1,220-1,640; inches:48,03 – 64,57/ Thickness: millimetres 2 - 3,2 – 4; inches:0,079 - 0,126 – 0,155. Glasal can be sold with detailed set-up explanation.

The large choice of available tones, plain or mottled, marbled or flax-fabric looking like, was equally a strong mean to diversify public and private decoration. Moreover, a ten years resistance and useful life warranty was a favouring bonus. Such a success obviously generated yearning and fierce competition between fibre-cement makers. Saint-Gobain, created "Murcolor", a prefabricated panel. Glaverbel invented his "Colorbel", basing his advertisement on the long life of colours. Everite, branch of Saint-Gobain, also tried his luck, without great success, with a product which they claimed identical to Glasal, but did not stay long on the market.

Such competitive campaigns brought the company to inform architects in strong terms. In her dissertation, Françoise Lampaert reported: *"Do not try to make a fake! Painted sheets, whatever the quality of their base, cannot offer the advantages of Glasal sheets with integrated colouration. Only Glasal, with its auto-cleaning surface can ensure different shades and tidy facades, with large durability... Glasal is both a product and a make, manufactured and sold by Fibrociment de Poissy...There is nothing like it!".*

06-023 Glasal-Elo Advertisement

Glasal came to last some fifty years, which is seen as something outstanding for such a material which largely overpassed the "thirteen glorious". The success was not only a French one, not even a European one. From the beginning in 1961, it was exported to Canada, Australia and Malaysia and won recognition in schools, gyms, hospitals, restaurants, cinemas, malls. It finally succeeded to be exported in over fifty countries[108].

Glasal was intimately linked to the happy post-war architecture at the time of asbestos. At the risk of anticipating on the next chapter, let us say that it succeeded to turn over a new life with the so called "new technologies", while maintaining its foremost qualities. As soon as 1986, the first asbestos-less panels were produced, and from 1994, asbestos was totally excluded from its components. It carried on its career up to the beginning of the 20th century, when new products for the building industry happened to kill it. *(It was replaced by a new Eternit product named "Equitone Picture")*

In 1967, in Rocky Point New-York, USA, a Doctors professional building was made almost entirely with asbestos-cement. The three-lingual "International asbestos-cement review AC 47" published pictures, plan and description quote:

06-024 Doctors professional building made almost entirely with asbestos-cement.

"The Rocky Point Professional Building is a small medical office building commissioned by a dentist to house his own clinic, two obstetricians and a paediatrician. It is designed on a minimum size plot along a two-lane highway. The two identical halves of the building have been offset to permit a small (soon-to-be) garden front and back and the reduction in length of the central public corridor. Full length windows serve office and reception spaces; the above eye level strip windows illuminate the high spaces of the operatory and examining rooms.

[108] Source « Usine d'Aujourd'hui »: Works and Project manager n° 66; Société d'Edition Professionnelles et Techniques, Paris 1961.

The asbestos-cement sheets are used as infill, set into redwood frames under the long windows; as continuous sheets, joined together by white aluminium battens for the facia; and as panels, covered at the joints with redwood battens on the sides."[109]

The same year 1967, in Sweden the same review confirmed the presence of asbestos-cement in ventilation ducts. These later were, at the time, becoming of general use in contemporary buildings.

In Australia, still in that same trilingual review, we discover that a local company designed a new type of formwork, quote:

"The formwork was used for the first time in the construction of the N.S.W.[110] *Housing Commission's impressive height storey block of 226 flats now nearing completion in Mascot, N.S.W. The contractors were given special approval to use an entirely new*

06-025 Asbestos cement formwork in Australia International A C review 47

[109] International asbestos-cement review AC 47. Estimated date: July 1967.

[110] N.S.W.: New South Wales, free state of Australia, capital Sydney.

type of formwork designed by them, which has permitted speedy erection, ease in dismantling and has resulted in very straight and upright walls. It was made possible by the use of thick compressed asbestos-cement sheets".

"Formwork frames were made to suit the internal wall sizes, the smallest being 12'0" x 8'0", and the largest 32'0" x 8'0", which latter weigh 35 cwt. It was found that each frame could be used 36 times on an average. To facilitate dismantling, the asbestos-cement sheets were treated with a form of oil prior to the assembly of the formwork in final position. The system permits the reinforcement rods to be placed in position on the formwork frames whilst on the ground spacers are positioned in the frames to ensure the correct 6" wall thickness".

Finally, the review offers an interesting description of the British Pavilion at EXPO 67 in Montreal Quebec, Canada: *"...it incorporates asbestos-cement materials for the entire vertical surfaces of the structures which comprise exhibition halls in two blocks, dominated by a 200 ft tower.*

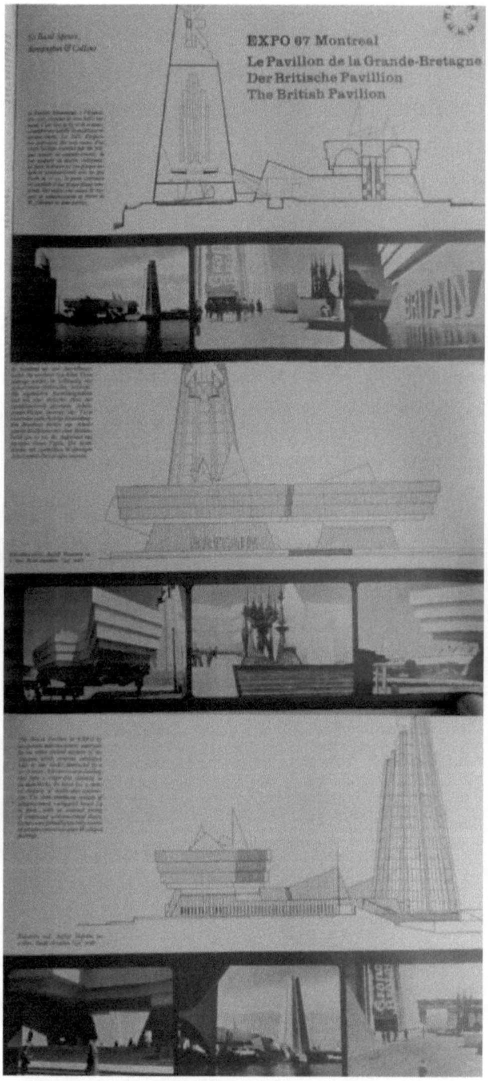

06-026 Expo 67 Montreal- the British Pavilion AC 47

Asbestos-cement cladding tiles form a single skin cladding in the main blocks; the tower has a vertical cladding of double-skin construction. The inner membrane consists of asbestos-cement corrugated board (4 in. pitch), with an external facing of compressed asbestos-cement sheets. Corners were formed by specially moulded asbestos-cement two-piece W-shaped flashing"

. On another note, in 1973, the Department of Wall Art in the Ideal Home Exhibition in Paris, suggested to a group of artists, to work on a project with the following requirements:

- One theme: wall animation, - one technic: sand-blast engraving,
- One material: Glasal, capable of assuming works durability both inside and outside, choosing one shade among thirty-five offered by
Eternit.

"Result: uncommon, coloured, strong works, have a look wrote Guillaume Gillet[111].

06-027 Wall Art Ideal Home Exhibition Paris 1973

[111] « Le Fibrociment dans l'Art » Elo 1973, already quoted.

In 1978, Josep Guinovart, Catalan painter, engraver and draftsman, discovered in his turn, the possibilities offered by fresh asbestos-cement paste. In order to work as he wished, he went to Uralita in Cerdanyola, just like Oscar Lena years earlier. A video was shot on the occasion. Rehabilitated in 2009, it shows not only how the artist operated, but also how a sheet machine of the 1950s was working. The video[112] also helps understand how easy moulding fresh paste was. A visit to the website will definitely interest the reader eager to clarify his sight of a sheet machine and the birth of an asbestos-cement sheet such as explained in chapter I.

06-028 Josep Guimovart Cerdanyola del Valles Cataluña España, around 1950

[112] Click on the link Josep Guinovart, the video is somewhat long to open, just wait, it is worthwhile.

A South-American engineer, friend of mine, who has been working for years in the fibre-cement world in which he is an expert, tells us:

"Fresh paste has always been a material which, thanks to its extreme malleability and its moulding facilities, was suitable to multiple 'sui generis' *applications. One could mould anything, as much as you wanted, without any problem".*

"In South and Central America, fresh paste was used to mould joints on roofs with strange and complicated shapes. With fresh paste we would make water- tanks from 200 up to 2000 litres. We could also make window boxes in at least thirty different sizes and shapes.

"In Bolivia, we had applications such as rehabilitating church cupolas, even on cathedrals. We also made satellite dishes up to 5 metres (200in) in diameter. Metal meshes were embedded between mono-layers. We also made small advertising items, which are decorative models adorned and painted by hand.

"I remember that indeed, we would send semi-liquid fresh paste to small companies, who poured it into moulds to make statues or any kind of decorative gadgets used in feast days celebrations, etc.

"Unfortunately, I don't have any picture anymore that was so long ago..."

06-029 Typical Brazilian Church by anonymous artist. Jacques Roulland private collection.

In Curitiba, Brazil, an unknown artist reproduced a model of a typical Brazilian church. The manager of the Brasilit factory of the town offered me one of them on the opportunity of a trade visit.

In 2018, an Indonesian company, with representatives in Europe, is offering fibre-cement elephants, asbestos-less, whose aspect and touch actually remind the skin of the mythic beast. One of them on the cover of the present book.

In the first decades of the 21st century, artists and designers do not hesitate to grab a bunch of fresh fibre-cement, which just fell off the machine, to conceive decorative works, as can be seen in the following pictures and on the following website Journal du Design.

•

06-030 Trash cube by Nicolas Le Morgne. *06-031 Soft light by Rainer Mutsch*

Both out of Eternit Design Journal

06-034 Changchun Century Movie park, China 06-033 Underground South Station in Shanghal

Three works out of different Eternit websites

06-037. Two Chinese works made with New tech fibre-cement.

Upwards: 70 000 sqm GRC (Glass Reinforced Cement) at Changchun Century Movie Park.

Downwards: Shanghai Underground South Station, with fibre-cement covered walls.

The quantity of artistic works made with fibre-cement is so important that it is absolutely impossible to quote them all. An entire book could well be dedicated to the subject.

VII.. Wind of change

"A total reconversion for Eternit and a careful one for Everite. After asbestos prohibition, two former manufacturers decide quite different reconversion strategies." L'Usine Nouvelle, October 6 1996.[113]

Duly summarised by l'Usine Nouvelle: the future of fibre-cement manufacturers, fiercely stroke by asbestos prohibition, let managers wondering... Discussions were underway between them. One point was obvious for everyone: Production cost was going to increase. Investments were required, without being sure that they would be the right ones. They meant new technics, new machineries and new raw material not yet one hundred percent tested. What was less sure, was the behaviour of customers facing high and inevitable price rising. Moreover, new products, were appearing and began to reduce market shares: roofing and wall enamelled steel sections for outer and inner use; PVC and PE-HD[114] for sewer and any pipe system required for our well-living.

While the new products expect to be totally available and accepted by everyone, l'Usine Nouvelle of the same time also reported: *"In order to prevent workers of the building industry from being exposed to floating asbestos fibres, manufacturers offer 'ready -to-use' products. They are originally cut or drilled in the factory, where workstations are fitted with protecting device. For instance: corrugated sheets due to roofs are delivered with their angles already cut. As to front walls, or man-holes, Eternit offers 'custom-made'. Architects supply us with their plans and the number of*

[113] L'Usine Nouvelle: Famous weekly French review, read by industrial managers in France and French speaking countries.
[114] PVC: polyvinyl chloride. PE-HD: High Density Polyethylene.

needed sheets or panels are calculated thanks to our special software, explains Henry de Belsunce marketing manager".

We saw that in France, Everite gave up with the conventional system at great expense with his Novatech, in the transformed Descartes factory and the new one in Dunkerque, exploiting the Wellcrete technology. The new essential material was glass-fibre, which could be used both for flat and corrugated sheets. We also noted that there was no surprise as Everite was a branch of Saint-Gobain, number one in glass industry. Finally, difficulty and cost soon destroyed Novatech in spite of Wellcrete sheets to resist impacts.
Eternit France, helped by the experience of its European sister companies, who were already "asbestos-less-converted", some of them for almost ten years, made another choice. They replaced asbestos by PVA[115] and wood pulp, and at the same time adapted the Hatschek technology which had proved so efficient for decades.
The option not only reduced immediate investments, but it also offered two more advantages:
At least for a while it allowed to maintain with and without asbestos technologies. That meant: let time to dominate the new technics, producing small quantities of asbestos-less slates. They could easily be exported to UK and North European countries where asbestos was already banned for several years, being thus widely ahead of France. This is how, as soon as 1996, the Thiant factory was producing 1.7 million square metres of asbestos-free sheets, that is 17% of its yearly production. Other French factories of the group soon operated equally, permitting that 10% of the global production be released from the now satanized fibre.
One must not burry one's face in the sand. Transition was going to meet technical problems and heavy investments, as testified by people in charge at the time that I was lucky enough to meet and interview. The same article in "l'Usine Nouvelle" already quoted several hundred million of French Francs to be invested, without specifying if it dealt with the only Thiant manufacture or with various brownfields of the company. I like to believe that the amount reverted to a globalized operation. Let us remember the forecast expressed in the Kuala-Lumpur conference abundantly quoted in the Crisis chapter.

---

[115] PVA: Polyvinyl acetate.

Another sight of things was suggested by Syro - a Swiss manufacture of fibre cement machines- that offered complete manufacturing systems, in various sizes, for asbestos free fibre-cement process. In its presentation page, Syro suggested a large list of questions met by any industrialist ready to turn from asbestos to non-asbestos, without losing sight of the promising market that he foresaw:

"***Questions to answer and decisions to consider:***

Modify the existing production line or invest in a new one?
Which machines needed to be modified?
Which devices need to be modified?
Which production flows need to be changed?
What additional devices are needed to be installed?
What ingredients are to be used?
What are the costs involved?
What is the time frame needed?
What is the most efficient procedure going forward?

The same company was so confident in the future of the products getting out of its machines that it did not hesitate to offer quite unusual width. Just have a look:
On his Website:

« Range and Types

Single width/single, double, or triple length, un-pressed, air cured?
Single width/single, double, or triple length, pressed, air cured?
Single width/single, double, or triple length, un-pressed, autoclaved?
Single width/single, double, or triple length, pressed, autoclaved?
Double width/triple length, un-pressed, air cured?
Double width/triple length, pressed, air cured?
Double width/triple length, un-pressed, autoclaved?
Double width/triple length, pressed, autoclaved?

In spite of a Swiss origin and of soft prices, thanks to an India based production, Syro was acquired by another Swiss group in 2016 and seems not to have sold any machine during the last years.

#fibre(S)cement.com

In this first quarter of the 21st century, Eternit, ever present, offers a smart option with a proud Hatschek machine settled in Saint Gregoire near Rennes, Britany. One-hundred percent automated, the machine is mixing: cement, calcium carbonate, silicon smoke, pure and recycle pulp, and PVA. In the end, very fine slates come out, as can be seen on the video viewable via the above weblink or QR code. Product and working conditions are very strictly observed during the entire manufacturing process. To improve the brand image, the "ersatz look", so poorly looked upon by roofing professionals, had to be absolutely eliminated. Once the useful life time of the colour was solved, the company fought to improve the relief in order to make the slates look like natural ones. Upgrading of the finishing stage on the machine allowed to reach the expected goal. Yet, rims also had to look like those of original slates, which was not so easy to achieve. Among employees working at the French Eternit Research Department, one was a former roofer. In 1996, he had the idea of making his slate cutter go wrong in order to get a jagged cut instead of the usual clean one. He suggested his idea as a project development. It was acknowledged and it turned into a new "chipped-slate".

07-001 Eternit NT tile produced in Saint Grégoire Plant near Rennes France, Eternit review: " Solutions for your buildings"

07-002 /07-003 Modern roofing with asbestos free fibre-cement

Slates were not the only products to come back on the market thanks to "new technology": Corrugated sheets also came to a new life, especially as underlayer for Roman tiles. They make the setting easier and highly improve the roof waterproofness using the "tegula/embrix" technic already quoted in chapter IV.

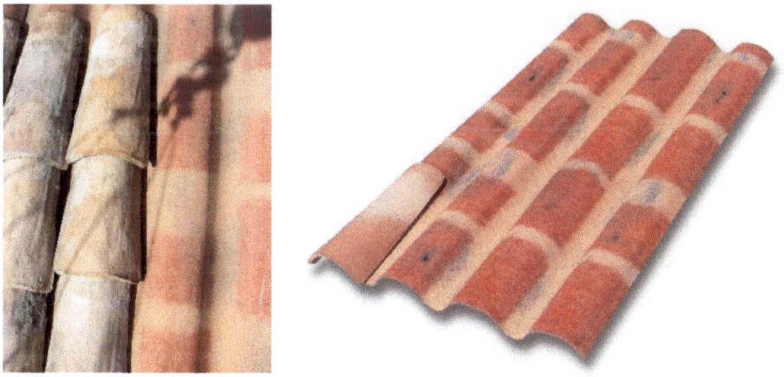

07-004/07-005 Tiles and corrugated under sheets by Eternit.
Eternit Review: Solutions for your buildings

On the scheme below, one can easily note that a modern Hatcheck machine is quite similar, though much more efficient to that of the beginning of the previous century, and that the "ribbon-sender" already mentioned, is still present on a contemporary machine.

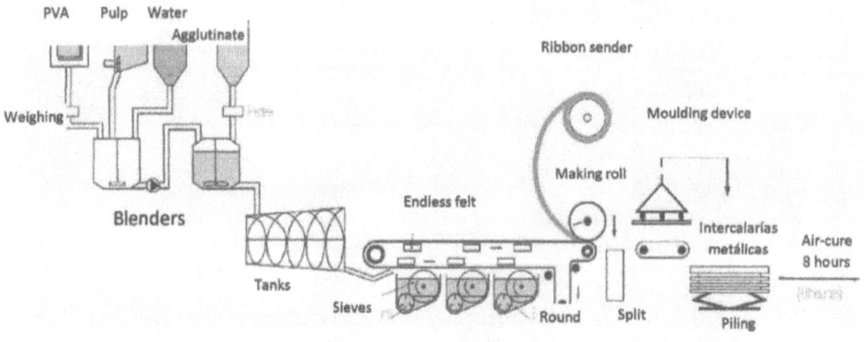

07-006 Scheme of a modern Hatcheck sheet machine.

One also notes the likeness of the main parts with those of ancient machines: silos for paste mixing, round sieves, suction boxes, felt, making-roll, but now-on two more devices appear: one is meant for the supply of PVA (mostly provided by Japan) and one to supply pulp. The device supplying asbestos has, off course, disappeared.
Among Eternit products closely linked to the happy architecture of the post war years in Europe, one must consider Glasal, already mentioned in the previous chapter. The huge success met by these panels at the time of asbestos-cement, entirely justified research and investment to offer it a new life in asbestos released technologies. This is how Glasal gloriously reappeared on the market, even before the general asbestos ban. As soon as 1986, the first asbestos-free Glasal panels were manufactured, and from 1994 all of them were such.
Eternit gives us some details regarding the ingredients entering in the new fibre-cement. On the way, they do not forget to praise its qualities.

https://www.eternit.co.uk/en-gb/

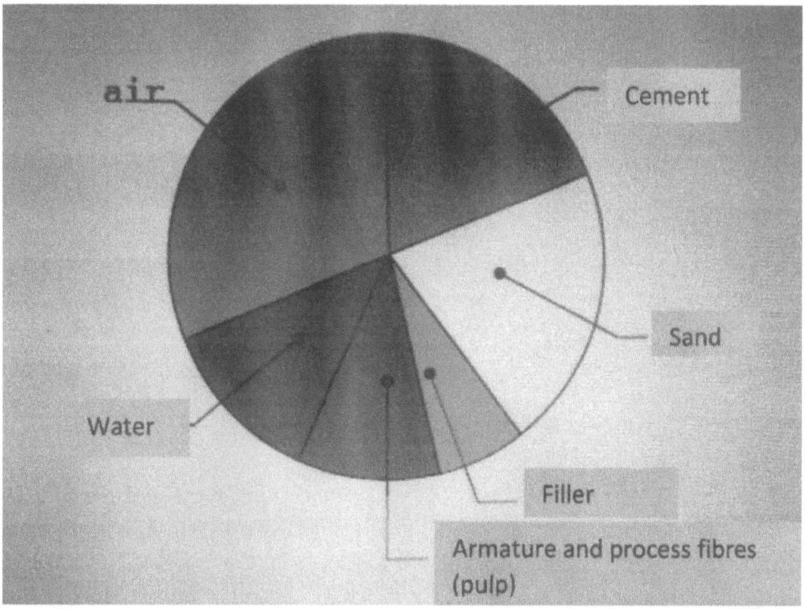

07-007 Eternit new Fibre Cement NT composition/ One of Eternit websites

Then, they add: *"Asbestos is a bio-persistent chemical structure (silicate). In the opposite, most important fibres now used by Eternit in the case of fibre-cement, have an organic chemical structure, i.e. are entirely different. Pulp is made of carbohydrate and the chemical structure of PVA is close to that of polyethylene and of polypropylene (daily used plastic). Not any serious scientific institution (IARC, MAK Komission, HSE, NTP) ever classified these organic fibres as carcinogenic or even, possibly carcinogenic.*

Every above organism enjoys a worldwide scientific authority. They all are specialized in the field of carcinogenic characteristic of substances and are in charge of defining the thresholds of these substances regarding exposure of people on work-sites (MAK Komission and HSE).

These substances are commonly used in industries for more than 50 years. Fibres classification by scientific national and international organism is based on lab tests and epidemiologic studies.

ARC (International Research Agency, linked to WHO) in charge of research against cancer; MAK Komission: world known College of German toxicology experts; HSE: Health and Safety Executive, counsellor by the British Government; NTP: National Toxicology Program, counsellor by the US Government".

07-008 Sights of fibre-cement panels used in motorways tunnels

One particular point will be of interest for car drivers travelling through motorway tunnels...Promat, one of ETEX[116] branch, presents fireproof panels called Glasal T, which are used to recover the bottom vertical walls of numerous highway tunnels.

According to the manufacturer, more than 250 tunnels in the world are doted of such panels. Such is the case of « Tunnel du Mont Blanc » [linking Italy and France under the Alpes], which was rehabilitated in 2002, after the dreadful fire that killed 39 people in 1999. Not only is this material fireproof, but it is said to soften light change to the benefit of drivers and passengers. In addition, walls polluted by car exhaust are easy to clean.

[116] If you ever get lost with company names, just remember that ETEX is the 21st century's name of the former Eternit.

Of course, asbestos-released products are not just Glasal, but quoting them all would consist in writing Etex catalogue, which is not our purpose. Yet, I cannot resist to my wish of showing you an architectural work which I consider as actually worth to be known, located somewhere on the North See coast, in France or in Belgium.

07-009 Strangely, in spite of a long search, no one in Etex could tell me where the buildings are, neither who took the picture. If any reader can recognize the place, please be so kind and let me know. (jcq.roulland@gmail.com)

So many are the possible mix offered by fibre-cement that would make one think of a "Prevert inventory"[117].

Yet, we cannot avoid to mention another product, closely related to climate warming, which might be of interest for many countries; we are talking of "Verdura sheets" which support a multi-layer plant-covered complex to be set on building roofs. *(Also manufactured by Etex, but made of polypropylene and consequently off-topic).* Here under summary:

[117] Prevert is a famous French poet of mid-20th century, who wrote an also famous inventory mixing things which had no common points.

Verdura sheet, is a profiled cement reinforced, rot-resistant, flame and fire-proof sheet, size 1.20 x 0.96 m (47in x 38in).
Linked to Hydropacks *(Etex licence)* Verdura sheets enable to turn roofs into vegetal areas. A very up to date idea.
Here again, a visit to the above website will offer interesting information though somewhat off our subject.

07-010 Verdura sheet

06-011 roof application of Verdura sheets.

By the way, let us revert to our given-up topic after our short digression of the last page.
In 2002, in France, Eternit, Branch of Etex Group since 1994, became Eternit SAS, and 8 years later was split in 3 companies:
-**CRI**, gathering the three French manufactures still alive.
-Haulchin, North of France, specialised in natural and colored corrugated sheets.
-Tersac, South-West of France, whose flagships are TN colored corrugated sheets. Cologri obtained by applying fresh pigment directly on the Hatschek machine and corrugated sheets used as "under-tiles" that can absolutely participate to the rehabilitation of old buildings, without any damage to their ancient look.
Saint-Grégoire, suburb of the Britany town of Rennes, specialized in the manufacture of slates. That will not surprise anyone aware of the slate ubiquity in this French region. Not only does this modern manufacture supplies its own area, but it also exports great part of its production to UK and North European countries.
-**Eternit Commercial**, gathering every member of the group dedicated to sales. -**ECCF,** which became an expert centre including every employment medium from the

three entities that emerged from the above separation. They are also dedicated to every other company belonging to Etex Group

07-012 Old South France building restored with Eternit NT corrugated sheets Eternit Review: Solutions for your buildings

A specialised prospective site, Fridonia, informs that the future of fibre-cement is globally very promising. One can notably read:

"Product Update:

Fibre-cement producers are continuing to create new ways for consumers to purchase their products more easily. James Hardie, for instance, has created an app that allows contractors to obtain measurements and a 3D rendering of a home just by taking pictures of the project. The company claims that the measurements are 95 percent accurate and prevents the need for measuring manually, saving contractor's time and money.

"Fiber cement siding will also see strong gains

Fiber cement siding is also expected to post solid demand growth through 2019. Primarily, because of gains in the US which accounted for nearly one-third of global fibre-cement siding demand in 2014. Fiber cement siding has become increasingly popular in the US as it can withstand harsh climates without rotting and is impervious to insects. Between 2004 and 2014, fibre-cement siding increased its market share by six percentage points. Gains going forward will be less dramatic, but still respectable.

"Study Coverage

This Freedonia industry study, Global Fibre-Cement, analyses the 26.3 million metric ton global fibre-cement industry. It presents historical surveys (2004, 2009 and 2014), forecasts (2019 and 2024) for the market (residential, non-residential, new construction, improvement and repair) and application (e.g., roofing, siding, moulding and trim) for 6 global regions and 19 major countries. The study also considers market environment factors, details industry structure, assessment of the company's market share and profiles 24 industry players, including Etex Group, James Hardie Industries and Nichiha."

The page also provides information about

"Asbestos-based products which are still popular in Brasil and India[118]

In 2014, roofing comprised the largest global market for fibre-cement products. Brasil and India are the largest markets for fibre-cement roofing products. Lower cost products (using asbestos as the fibre) are popular in these countries. However, growing health concerns about products made from asbestos will slow growth in these markets going forward. Fibre-cement roofing will lose share in Brasil and India to concrete and clay tile products, which are less expensive than asbestos-free fibre cement roofing. As a result, gains in demand for fibre-cement roofing are expected to be below average through 2019."

[118] Asbestos was in between banned in Brasil.

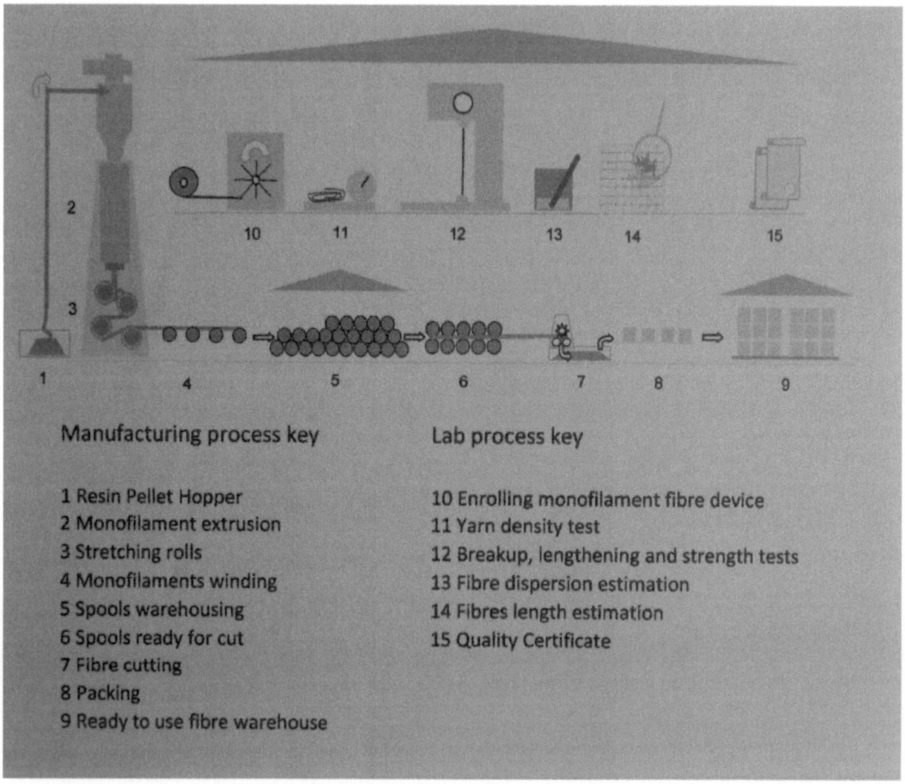

07-013. Diagram of process for PP fibre production. Fortaleza, Brasil.

My Mexican friend, who has lived both the death of asbestos and the renewal with new technology from within, explains in great detail the final steps of the struggle that made it possible to achieve the current situation. He insists on the fact that, at the turn of the millennium, under instigation of Saint-Gobain, Brasilit and Mexalit, each one in his country developed the production of substituting polymer fibres. The latter permitted to drastically reduce the impact of imports from China and Japan on production costs of the new fibre-cement, called FCNT, for Fibre-cement New Technology. He also offers us the above detailed diagram which sums-up the manufacturing process of the new fibre.

VIII Countries overview

"Exporting is the best way to protect one's company against the fossilization which has a dominant position on the home market and might lead to...A foreign customer is not different from a French one. You must know him; you must foresee his problems and gain his confidence". Roger Martin, "Patron de droit divin » already quoted.

If the above statement from a big French boss can perfectly be applied to the groups that he managed[119], it seems that it can also apply to most fibre-cement groups, whatever their country of origin, at least to those belonging to a free trade economy. This is confirmed by our visits to some countries where the industry has been successful or is still doing well.

Fibre-cement business, just like many other industries, is interested in the whole world and big industrial groups from at least three continents are very ambitious. All of them wish to develop their activity, especially in young countries with a growing prospect and a high birth rate, where needs in term of industrial buildings, offices and dwellings are gigantic.

From the 1960s onwards, creations, associations and mergers were all the rage. If I could write in the prologue that companies and human beings have very similar lives, I trust that I can add without much risk that the likeness still goes further. In the modern world, marriages lost their sacred and lasting charm, purchases and mergers abounded, but all the same did divorces and step-families created pains and misery among workers suffering of financial operations modestly named "restructuring". Some examples will appear in the next pages.

---

119 Roger Martin was President of Pam (Pont à Mousson) and of Saint-Gobain-Pont-à-Mousson after the merger of the two companies.

A for

Algeria

We already talked about three powerful and modern factories that were built in the country in 1974/1975 by a Belgian-French consortium. It seems that there is absolutely nothing left when I write the present book 40 years later. Every fibre-cement product offered on the Algerian market comes exclusively from Vietnam. Browsing the web, I was able to find an undated tender from the Ministry of Mining and Industry on behalf of the GICA Cement Group for a corrugated sheet factory. The invitation to tender seems to have remained unanswered.

Australia (and New Zealand)

In 1888, a young Scot named James Hardie, left his family and his native environment. He was looking for new opportunities and sailed to Melbourne. Good at tannery processes acquired in family business, he began by importing products required to develop his former business in his new country.

08-001 James Hardie's shop

In 1892, Andrew Reid, one of his friends from Glasgow, with whom he was still in touch, joined him. The two combined their knowledge, skills and aim in life.
During a journey to London in 1903, he heard about a new type of roof cover: fibre-cement which had just appeared on the continent and decided to import the new product to Australia.

Sales boomed, James retired in 1911 and sold half of his business to Andrew, but the name "James Hardie" remained and still remains with much excitement. The business grew dramatically and was family owned until 1995.

As soon as 1906, they began to export sheets to New-Zealand. In1925, they sent their own people to Wellington for a big roofing work. In 1936, they created Fibrolite S.A. and decided to settle a production factory with only one Hatcheck machine making both flat and corrugated sheets. Production started on October 8th 1938. A first major contract was won in 1940 for the Centennial Exhibition of New Zealand[120].

From 1940 to 1981, the New Zealand factory grew and modernised.

In 1951, James Hardie Industries entered the Australian Stock Exchange.

In 1981 occurred a merger between James Hardie, Philips and Impey, giving birth to James Hardie Impey, which developed different activities such as textiles, chemicals, machinery, wallpaper and building materials.

Around the 1980's, James Hardie began to produce "Fibre Reinforced Cement", currently called "FRC", made of a mixture of:

-cellulose fibres

-Portland cement

-sand and water

-sometimes bound chemicals are added to help the process and to give the product some distinctive features.

The company began to design and manufacture a wide range of fibre-cement products which benefited of their high resistance, versatility and useful life.

Thanks to technics born in Australia, James Hardie, who was then dedicated in the only fibre-cement industry, became one of the giants in that activity with factories in Australia, New Zealand, the United States, Chile and the Philippines. According to the James Hardie website, the company is active in the field "sustainable development":

-water and resource conservation

-energy consumption and management

-use of renewable and recyclable resources as raw materials

-avoidance of environmentally harmful raw materials

[120] Hundredth anniversary of the Waitangi treaty signed in 1840 between the British Crown and Maori Leaders. The date is considered as being the birthdate of New Zealand.

-waste minimisation through process recycling of materials
-pollution reduction and protection of the natural environment.

From 1990 onwards, due to the development of the group's activities in the USA, a new company with headquarters in the Netherlands was founded in 2001 for tax purposes, named: "James Hardie Industries NV". In 2004, by hard luck, a change in a tax treaty between USA and the Netherlands brought the group to search for another protecting country. After carefully checking the conditions offered by the different countries of EU James Hardie chose Ireland as the most favourable. Shareholders agreed 100%. Consequently, the new headquarters moved and a new company name was given: "James Hardie Industries Plc (JHI . Plc.).[121] [122]
Before we leave James Hardie and the wanderings of his headquarters, let us spend a few moments to have a look at his asbestos-less pipes.
In Europe and North America, asbestos-less pipes knew little achievement, even if their so-called flexibility was not without interest in case of works on hostile ground. The ISO 22306 standard was issued, but cancelled in 2007. It does seem that manufacturing was entirely given up since that time. Many specialists believe that producing high quality pipes with any other fibre than asbestos is totally impossible.
In the contrary, in Australia, VantagePipes , originally a branch of J.H., developed cellulose cement reinforced pipes as soon as 1984. It seems that they took the idea from the steam-cure process already in use since the 1950s for high pressure pipes. VantagePipes claimed exclusiveness and that more than 12,000 km were sold through Australia. The company offered us a nice scheme which reminded us of the Mazza pipe machines as already mentioned in previous pages, though including a few more details. In view to convince us of the quality of his reinforced cement pipes, VantagePipes invited us to visit the system installed in the biggest Brisbane's motorway tunnel[123]. This supports 40,000 vehicles a day and avoids the winding business centre of the town as well as 27 street lights, saving 30% in time.

The following map offers a clear sight about the growth of James Hardie from the time of the small Australian shop.

[121] Plc: Public Limited Company, quite frequent in Commonwealth countries.
[122] Information detailed above was extracted from various J.H. Websites.
[123] In December 2020, I am informed that companies now gave up manufacturing reinforced cement pipes as they are no longer competitive against concrete pipes and plastic pipes.

North America Fibre Cement

James Hardie is the largest lowest cost producer of fibre cement in North America

Plant Locations

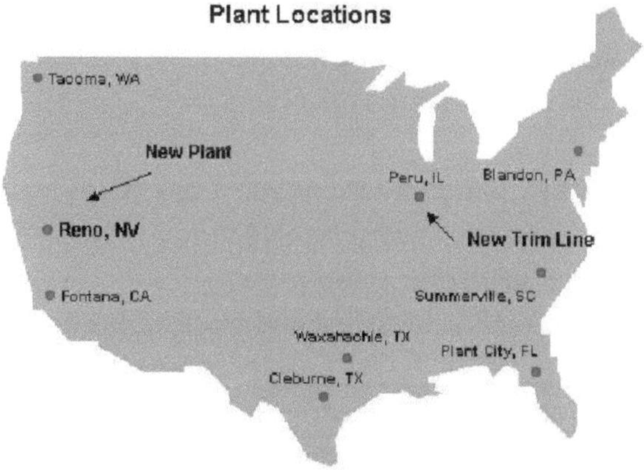

Plant Capacity

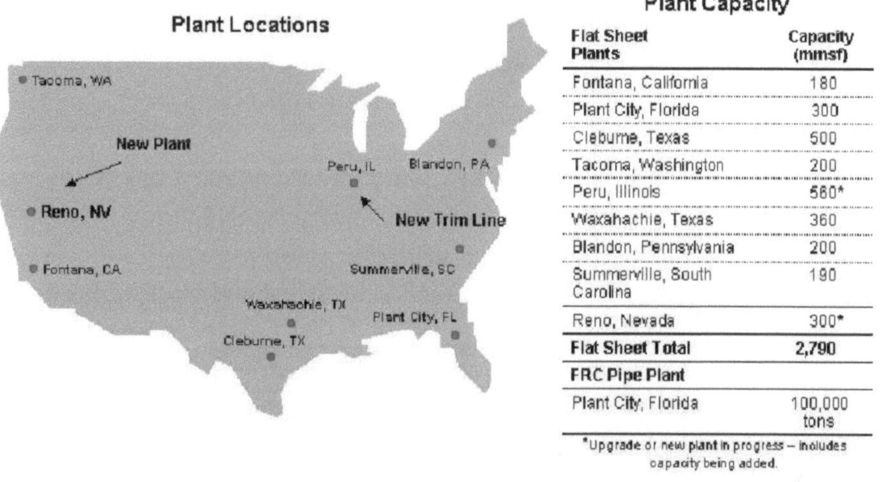

Flat Sheet Plants	Capacity (mmsf)
Fontana, California	180
Plant City, Florida	300
Cleburne, Texas	500
Tacoma, Washington	200
Peru, Illinois	560*
Waxahachie, Texas	360
Blandon, Pennsylvania	200
Summerville, South Carolina	190
Reno, Nevada	300*
Flat Sheet Total	**2,790**
FRC Pipe Plant	
Plant City, Florida	100,000 tons

*Upgrade or new plant in progress – includes capacity being added.

9

08-002 self-explanatory.

Austria

We are back to the country where Ludwig Hatschek invented fibre-cement in 1901 when he discovered the friendly alliance between asbestos and cement and produced his first sheets in the old plant of Kochmühle, near Vöcklabruk. For memory, he had purchased this factory in 1894 and its main activity consisted in producing rag-paper.

As soon as 1903, he registered Eternit as a make and the company became "Eternit Ludwig Hatschek AG". In a very short time, the invention spread over Europe and well beyond.

When he died on July 15th 1914, his son Hans, who was working in the company, provided continuity.
By 1930 Hans Hatschek began building the Vöcklabruk hospital, which was soon named after. In 1939 the town made Hans an honorary citizen and the street leading to the hospital became "Hans Hatschek Strasse".
On that same year 1939, the company's staff was reduced to 14 people due to the war and resulting mobilization. Soon after the end of WWII, the workforce has increased to 840 employees, as a result of rebuilding effort which gave the building industries so much to do.
A new factory was set in the Vienna suburbs.
From 1987, the Eternit Austria increased its search regarding substitute fibres and as soon as 1993, asbestos was entirely banned from every manufacturing. Seven years later, Eternit Austria, had become market leader for steep-roofs.
During the following years, the company went on investing and creating, particularly in ceiling panels.
In 2005, 80% of the shares were acquired by Cross Industries AG and in 2009 Bernhard Alpstaeg purchased 80% of Cross Industries AG through the "Fibre-Cem Holding" (Eternit Switzerland). The remaining 20% were purchased by the same one. In 2016, "The Fibre Cem Holding" became "Swiss Pearl Group AG".
For sure, the case is not the last monopoly game that we will meet along our "Countries Overview".
But, Eternit Austria, whatever the name, also offers us some intelligent architectural works: for more information, please visit the company's website.

08-003 Drawing of the original factory in Vöcklabruk.

Displayed creation on

Eternit Austria Website

B for

Belgium

Just like Austria which we just visited, Belgium was one of the European countries that soon intensely developed the manufacturing and use of asbestos-cement. If you ever travel through the country, just have a look upward to the nearby roofs and you will be convinced. The number of walls and roofs covered with the grey rhombus shaped Eternit slates cannot escape the sight of the attentive traveller.
In fact, Eternit appeared in Haren near Brussels as early as 1905, and we could see that a French industrialist came close to the Belgian field and gave birth to the French homonym.
In 1923, another manufacture was opened in Sint Niklaas named SVK *(Scheerders van Kerkhoven)*.
Other companies, perhaps less well known, also appeared. Let 's remember some of them like: Alfit, Modernit, Coverit...
The first point to keep in mind as to asbestos-cement in Belgium certainly is the leading international role plaid by Eternit Belgium, now ETEX Group.
The second one is the number of products invented and developed in the country, often in collaboration with other Eternit companies from nearby European countries such as France, Italy and Switzerland, who had a great worldwide success for years. The Belgium website materiauxdeconstructiondapres-guerre.be offers many examples. We will list a few of them:
-conventional flat panels just called Eternit;
-multi-coloured flat panels, grey, red, green or yellow, called Eflex;
-Acimex panels, with incorporated grains of sand in their surface;
-Exterelo panels for outside use;
-Elostone panels imitating natural stone;
-Romana and Gallia roof panels, wide, flat at a time light and thin;
-Ardex corrugated sheets, born in 1952, grey, pink, Havana or green, with a transparent synthetic resin layer;

-Menuiserite panels, including pulp fibres which would give them a pink or yellow soft touch. The insulation, fire and rot resistant as well as the water resistance gave them a very high popularity rate.
Once more, we would not come to an end with a list of products invented in the "low-countries" or the reader would suspect me of being in the pay of the company.

Bangladesh

In 1963, when the country was still East Pakistan, Team S.A. from Luxemburg participated and helped to create Aramit Ltd in Chittagong. At the time when I write, Aramit is the only manufacturer of "cement coated roofing materials, drainage pipes, ceiling insulation sheets" in Bangladesh as one can read on their Website.

Bolivia

Bolivia, together with Paraguay, is the only country of America without direct access to the ocean. Such a situation, added to an above sea level going from 90 to 6 000 meters *(98,4 to 6.667 yds)* and climate areas varying from tropical to polar, make a big handicap to easy economic growth. However, it did not prevent the country to develop his own fibre-cement industry, in the contrary, perhaps did it help! Dualist was founded in 1977, with participation of Eternit Switzerland and entered the multinational Elementia in 2006. Here are some examples of the light roofing products offered by Duralit on the local market.

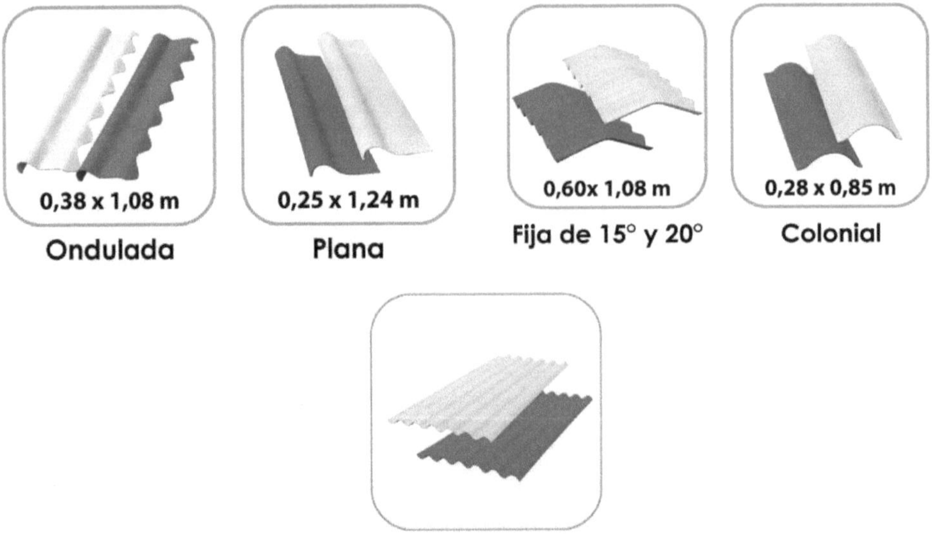

08-004 Duralit adds on their Website

Brazil

In 1937, the French PAM sent two of his engineers, both young but already experienced each one in his field, one in asbestos-cement, one in concrete. Their task consisted in starting a concrete pipe plant near Rio-Janeiro, where PAM had gained a contract for a submarine outfall[124] named "Ribeiro dos Lajes".
More or less at the same time, an Italian company, associated with local investors, was building an asbestos-cement factory, named Brasilit, in the suburbs of Sao-Paulo. This factory enjoyed a wide growth along the Second World War. Its success was mainly due to its flagship product of the time: water-tanks.
Another Italian firm also created a company "Civil Industria Artefatos de Cimento S.A" due to manufacture small asbestos-cement products. This one was also purchased by Eternit years later, in 1962.

---

[124] Submarine pipeline in coastal towns to move sewage as far from the shore as possible.

During the first month of WWII, both engineers from PAM, tried to sail back to France. Due to unexpected circumstances, they could not go further than Dakar *(France defeat against Germany May/June 1940...)* and decided to sail back to Rio Janeiro. Once there, they restarted the concrete pipe plant. Large profits have been made over the years which, due to the circumstances at the time, they could not transfer to their mother company in France.

On their side, the Italians in charge of Brasilit could not join their owners who could not leave Europe, more or less for the same reason. They met plenty of trouble which disturbed production. Consequently, they were trying to sell the factory...

Even without Internet or social networks, the guys in charge of concrete pipes and those in charge of asbestos-cement came to meet. The French, based on their profits made since the beginning of the war. purchased the young Brasilit. This is how, at the end of the war, PAM discovered that they owned an asbestos-cement factory near Sao-Paulo. The new owners, respected their well-known tradition and rapidly developed the factory, which soon counted with two sheet machines and two Magnani pipe machines, one making 4 meters (39,37in) long pipes, which at the time was quite significant. This last machine had a single particularity: once the pipes were freed from their mandrills, they would carry-on their way on a 60 metres long (65.6yds) subterranean chain. It seems that the adjective single is the right one, as according to PAM there never was any similar one in the world.

In 1938, the Brazilian geologist Hipolito Pujol, who was holding in his hand a piece of stone retrieved from a Bahian museum, discovered an asbestos mine with the help of a local prospector in a place known as Bom Jesus, in the village of Sao-Felix, Bahia State. The discovery gave birth soon after to SAMA *(Sociedade Anonima Mineraçào do Amianto),* which years later was managed by Brasilit belonging to PAM.

Asbestos-cement factories were multiplying. In 1946, the Brazilian Eternit Branch became: Eternit do Brasil Cimento Amianto. Part of the fund belonged to the Belgian Eternit[125]. The new firm settled a factory in Osasco/SP in order to produce flat and corrugated sheets. In 1947, Brasilit in turn settled a new factory in Esteio/RS.

[125] Roger Martin, « Patron de droit divin » page 242.

In 1948, Eternit do Brasil was listed on Sao-Paulo's Stock Exchange and built a new factory in Rio-Janeiro, when at the same time Brasilit was building one in Recife/PE, north of the country.

Resources at Bom Jesus mine were scarce and could not fit the needs of the new factories. PAM would send new engineers to manage the mine without much success. Production remained definitely insufficient.

All along the 1950's, PAM was both trying to develop his activities in Brasil and to improve his asbestos fibre production. A few years away, PAM sent two new young engineers to the country of samba. They both played an important part in the industrial development of which they were in charge.

In 1951, one of them, Charles Biaggi, entered the big Pont-à-Mousson family. The next year, just married for 3 months, he was invited to leave France and move to Brasil in view to assist the manager of a factory located in Senador Camara, near Rio Janeiro. In fact, the deal consisted in putting in order a concrete factory that was facing internal problems. Before leaving, he wanted to know about his future working condition and asked the chief of staff. This one did not know anything about, in turn he questioned the general manager by phone and got the following answer: *"When a young man is invited by PAM to go, he goes if he likes, and if he does not like, he stays"*.

Ignoring that Portuguese was the language used in Brazil, and willing to prepare his arrival, the young man began to learn Spanish while he was having a training course in the factory of Everite Situbé near Bordeaux. He was there to learn about asbestos-cement technology. Wondering about the local currency, he found that bankers in Pont-à-Mousson had no idea. They did not even know the word Cruzeiro![126]. Luckily, the young bride agreed to the adventure to come, and the couple flew the long trip with stops in Madrid, Dakar and [127]Natal before reaching Rio de Janeiro.

Before leaving, Charles was told that he would be assigned a small car. There was no company car in arrival. As for the dwelling, it was not much better and the couple had to stay in the hotel for a while.

[126] Cruzeiro: former name of Brazilian currency replaced by "real" in 1993.

[127] During WWII, Brazil allowed Brits and Americans to use their air-force base of Natal RG to grant their planes fly between USA and Ouest-Africa. After WWII, the air-base was also used by private airlines to fly between Europe and Latin-America until the arrival of long-haul jets.

The task in Senador Camara did not last long. The factory was making 3.10m (120in) concrete pipes intended for a submarine outfall that the French *Spies Batignolles* was setting in Guanaraba Bay near Rio-de-Janeiro. Charles Biaggi was soon sent to Esteio, a suburb of Porto Alegre - Rio Grande Do Sul - and soon after to Recife. He spent most of his working life in Brazil and stayed there for about thirty years. His long stay gave him the opportunity to participate intensively in the development of Brasilit. The fact that PAM merged with Saint-Gobain in 1970 did not prevent him from working his way up to the general manager of Brasilit.

But let us go back to the beginning of the couple in Brazil. Thanks to Charles' career advancement, they were soon granted a company accommodation. The house was located in Santo-Andre, an industrial town in the suburb of Sao-Paulo. It was quite comfortable, but surrounded by open air gutters and wooden shacks where local workers lived. Access was by dirt roads, except one which was tarmacked a few days before the elections and un-tarmacked a few days later. At the time, the local governor had his own advertising slogan: *"I steal, but I do"*.

Typhus was quite common but most Brazilians refused vaccination. In order to set an example, the young French engineer did not hesitate to be inoculated. Perhaps due to a lack of care in sanitisation, he almost lost one arm.

The two were rather welcome by their neighbours. One must admit that the Corsican origin of the man, and his knowledge of the language facilitated his approach of Portuguese, which was a good help to penetrate a quite different world from the one he used to know in his native land.

The town where they lived, Esteio, counted some 25,000 inhabitants and some 50 brothels. Common use was that men spend the end of the day, at least as long as the week's wage was not run out, in one of these poorly recommended establishments. A neighbour lady, married with a Brazilian guy, was very worried by her husband's evening trips. The French couple fought very hard to help change people's custom.

During the first half of the 20th century, Brazil waged several wars with his neighbours. Soldiers got used to drink *mate*, also called *terere.* They would drink it cold, as experience had told them that fire was too good a mark to get shot. Thus, drinking *mate* soon became a rule in Brazil. Neighbours living in the nearby shacks would often come to see the young French couple, now parents of two children. They liked to sup mate in a *chimarrão* and have a little fun.

Among the surprising practices that took place in the village of Esteio in the mid-20th century, one could note when the governor's election was coming soon, that around 500 kilometres of pipes could be ordered, even if not any major work was planned and these pipes had no likely use.

08-005 chimarrão

During the 1950/1960 decade, Eternit do Brasil organised a wide sales network to meet the requirements of the country, *(up to that time, the plants managed their sales in their respective area)*. They also increased their production of tiles, water tanks, pipes and flat sheets all over the country.

These years were happy times for asbestos-cement makers, Brasilit's production rose from 40,000 tons to 700,000 tons a year.

Asbestos needs were endlessly increasing in Brazil as well as in other countries. PAM was well aware of the fact and actively continued his quest for a safer own supply.

Roger Martin, already quoted, reminds that he was making numerous trips to South Africa, one of the biggest producers of the so coveted raw material. In 1953, during one of his journeys, he met a young engineer who was going to play an important game regarding the future of PAM and Saint-Gobain in Brazil. His history, deeply linked with that of SAMA *(Sociedade Mineraçao do Amianto)* is worth a few pages.

In 1939, the Milewski family was fleeing from Poland due to the German invasion. After a long haul through Turkey and Palestine[128], they reached Northern-Rhodesia[129], which was then a British colony.

Among the family members was a ten year old boy, Joseph. He would follow his secondary in different improvised Polish schools during WWII. Then he went to the University in Johannesburg, South-Africa. Once grown-up, he became geologist and started his professional life in the neighbour country of South-Rhodesia[130], an important asbestos producer. Soon he worked in asbestos trade in South-Africa. By chance, a young French girl had left her country for private reasons. They met, she

[128] British colony at the time.
[129] Zambia since 1964.
[130] Now Zimbabwe.

became his wife and shared their long, adventurous life over the years. In the course of his job he met Roger Martin, *(then high executive and future President of Pam)* who was always in search of asbestos supply. Joseph told Roger Martin his wish to go to France. Roger Martin took him at his word and offered him a six-month training job. Then, after an unforgettable air journey on board of an old DC3 surviving WWII and with numerous overnight stops, the couple arrived in France. The training began in Dammarie-les-Lys, near Melun, some 50 kilometres south of Paris, but turned into a four-year stay, part of it in Pam's Paris office. There, he began to get bored and thought to head over to new horizons more in line with his geologist basic schooling. Still resourceful, he obtained that his manager entrust him with an important task perfectly fitting his risky taste: locating a long searched, supposed, big asbestos mine in Brazil.

After a South-African detour to meet his family and have some rest, he landed in Sao-Paulo in 1958, with his wife and their two kids, the younger one being a few months old. He was there to get in charge of the already mentioned SAMA, which was running the Bom Jesus (*also called São Felix*) mine in Bahia, and to undertake the research of the nowhere to be found asbestos mine. His arrival as a "boss" did not delight the man in charge. In spite of the large distance with the head office in São Paulo where he was based, he would spend a few days at the mine every month and the situation soon improved, thanks to the tempered character of the newly arrived chief.

In this state of Bahia, our friend was struck by poverty, by men who sat outside all their time, dressed with a kind of pyjama, and by a general lack of care which enabled pigs to shuffle among people dinning in restaurants...Nearby the mine, his spouse would try to educate some miner's wives. She would teach them basic hygiene concepts such as boil feeding-bottles, chase flies from babies or move away concubines...Of course, drinking water was unknown. The mine manager ordered a dump to be built, giving birth to a lake where everyone could help oneself. A case of small-pox appeared among kids. Joseph Milewski isolated the family and urgently sent for vaccines. As kids never had any contact with the virus, they developed impressive pimples which made mothers crazy. They blamed the director for trying to poison their children and threatened him with severe reprisal.

The own children of the couple would play on slag heaps and they would come back home covered with fibres.[131] No one had any idea about the danger. Asbestosis was known, but one thought that it was limited to the workers of the mine.

Managing the mine of Bom Jesus was not the main task, nor the most interesting, for the geologist. He undertook prospecting and improved efficiency studying aerial photos obtained from a specialised company. Ultrabasic uplands from Goiás were famous for their nickel deposit up to Niquelandia, but not further away. Then he undertook his own aerial research using *"taco-tecos"*, such was the nickname of small mono-engine little airplanes which could fly over the region at low level, and permitted to locate areas with the promising colour of serpentine, the so coveted mineral.

After four years of hard work, a defined area was finally located as being the most likely to confirm the year-long dreams held in Parisian offices. A tough task remained: go and see on site, available roads of any kind also belonging to his dream. In 1962, a small expedition, consisting of two people on board a Jeep made way from Goiania and reached the small village of Campinaçu. Some 120 kilometres of poor track had been needed once left the Brasilia-Belem main road. In the evening, Eternit's prospector also arrived there as he was proceeding to the same research with the same purpose: be the first one on site in order to obtain the concession from local authorities. The guide accompanying our friend had a good idea. He convinced the competitor that going further might be quite dangerous... The village cacique, a future Goiás MP, entrusted Joseph with Pedro Paraná, his assistant. The next early morning, without a guide but with horses and supplies, they both left towards the site located from the sky. PAM's prospector had a limited confidence in his fellow traveller, during the night stop he slept with his gun at hand, ready to deal with any unforeseen event.

Joseph Milewski and Pedro Paraná rode some 70 kilometres in the *Cerrado*[132] during a day and a half, towards the *Garimpo.* The *Garimpo* is a Brazilian savanna scattered with small and stunted trees protected from April to September by total dryness with very thick bark. The land there is also covered by a dense grass of little feeding value,

---

[131] More than fifty years later, parents and children were still in perfect health. But finally, Joseph Milewski died from asbestosis in 2019.

[132] Cerrado: Brazilian savanna scattered with small and stunted trees protected from April to September by total dryness with very thick bark. The land there is also covered by a dense grass of little feeding value, explaining the poor level in term of fauna, compared with that of the African savanna.

explaining the poor level in term of fauna, compared with that of the African savanna. In the contrary, on the slope of "Serra de Cana Brava" which later on gave its name to the mine, flourishes a dense tropical forest with a rich animal life (*jaguars, wild pigs, some monkeys...*).

On April 28th 1962, at the end of their ride, they reached a place where they met prospectors searching to sell "*pedra-cabedula*" (haired-stone). They knew that it had some value. Joseph Milewsky made them understand that extracting the fibres out of the stones required expensive machinery and skilled workforce. He asked the *garimperos*[133] to clear an 800 x 30 meters area in order to prepare a basic landing strip. He paid them half the estimated agreed price and promised to pay the other half when he would be back within a fortnight. Moreover, he promised to offer them work for setting up and operating the future mine. After a ten hours ride on a mule, he joined the airstrip of *Anterrão* on the other side of the *Rio Tocantins*. He asked a small plane to come. Then he flew back to Goiania where he had departed from one week earlier.

From that moment, SAMA could start to apply federal authorities for the first right to search the mineral. A second one was solicitated at the end of the year.

On May 20th he was back on a mono engine Cessna 195 airplane flown by Moacir Mendonça, who, by luck, did not fear anything. The plane was also bringing food and an easily demountable drill. The very rude landing strip had been prepared by one of the prospectors, Alexandre Alves Pacheco with help of manual tools, such as a scythe, axes and hoes. The ground had been more or less smoothened with a heavy tree trunk harnessed to a pair of oxen. Joseph Milewski kept his word, as promised he paid the second half to the *Garimperos.* That made him gain their confidence and consideration.

Before starting any industrial operation, one had to organise a base camp including a small shelter, some 9 x 5 meters, in which a kind of office and a bed-room could be set up. A pathway to the Bonito River was also required in view to warrant water supply. A clear area, some fifty yards wide around the outcrop area, also had to be prepared to facilitate the settlement. Even doors, windows, tables and chairs had to me made on site. One also had to test extracted stones to confirm their value. Joseph obtained from a local farmer near the discovery area, access to a mortar and a pestle

[133] Garimperos: inhabitants of the Garimpo.

to break the stone samples and extract the asbestos fibres. He even came to borrow the house oven to dry the samples faster.

Obtaining exploitation rights appeared much more difficult than expected. The Goiás State Government considered that the deposit was his and that consequently he deserved 20 to 30 % of the turnover. That was obviously impossible. After a very bitter bargaining, 5% were considered as acceptable by both parties. Financial problems were not just with the local government request. Eternit also claimed to be a natural beneficiary as they had discovered another nearby deposit in a place named Niquelândia, as soon as 1961. In 1964, talks about SAMA began between PAM and Eternit do Brazil (*supervised by Eternit Switzerland)*. The two companies finally agreed to share SAMA 50/50.

Obtaining the right to search from the Mine Ministry required talks with local authorities that lasted three years, from 1962 to 1965. Then, two more years were still needed before publication of the Miner Exploitation Decree. Exploitation could actually start in 1967, when official agreements were reached between PAM-Brasilit and Eternit do Brazil.

Eternit brought one million US dollars in the bride's basket. Strangely, Pam who had spent a lot in prospection did not wish to invest any more. PAM limited his share to a few machines transferred from the then closed Bom Jesus mine and to some sixty experienced people also coming from the old mine.

Transport of the machines was also a kind of adventure due to the lack of roads between the two places. Trucks had to drive southwards by the coastal road from Salvador to Rio, then north/northwest, via Belo-Horizonte, Brasilia, Annapolis, Uraçu and Campinaçu, through roads that were often in very poor condition. Incidents and delay multiplied. Small drilling devices, to extract bore cores had to be air shipped by small airplanes which sometimes also brought explosives. While expecting the right to exploit, a research workshop was even built in order to test samples of fibre before industrial trials in Brasilit and Eternit factories.

Finally, the start took place and production began...[134]

[134] Two fine books by Renato Ivo Pamplona were published by SAMA around 2002.It might be quite difficult to get them. They tell details of the history of Cana Brava mine, together with lots of pictures. In English: "SAMA MINE 40 YEARS OF HISTORY Minaçu-Goiás -Brazil", in Portuguese: "SAMA 40 ANOS Minaçu- Goiás".

1965: *General view of the region and Cana Brava mountain. At the foot of the mountain to the left is the deposit of the* pedra cabeluda.

1962: *The house of cowhand Darcy Lopes Martins, at Fazenda Velha.*

1963: *Zacarias Pimentel (left) and Álvaro Lima (right), a Fazenda Velha.*

1963: *Tents used by SAMA at the beginning of the discovery of the*

08-006 Cana Brava out of "SAMA MINE" by Renato Ivo Pamplona

Roger Martin, already quoted, tells us in his memories that he visited the mine in 1968 and also adds:" *"The quality of the mineral is doubtless and works are going on at full speed. The first tons of fibres, some 150, had been shipped in July 1967"*.
During the wet season, roads were impracticable to trucks and sometimes even to four-wheel drive Jeeps. At the beginning, buildings were very basic.
Every infrastructure had to be built, homes, damp, roads, schools, hospital, church...Such important works required years and could be performed according to profits resulting from the mine activity, when mills and crushers got bigger, being thus more efficient and allowing production to grow. Most of the profits generated by the mine were invested in the needs of the mine itself and in infrastructure required by the growing local population. This is how the NEW MINAÇU appeared. Does it not remind us of what we read about ASBESTOS CITY in Canada?

In 1969, Can Brava mine produced 1,500 metric tons.

In 1974, it reached 11,000 metric tons and in 1976 the goal was 230,000 tons per year. The goal was only reached in 1999.

In 1997, PAM, which has since become part of Saint-Gobain, withdrew and sold his 23% shares to Eternit, adding it to its own 28% and becoming a majority shareholder. In 2008, though Brasilit had given up the use of asbestos since several years, production came up to 300,000 metric tons, mostly due to exportation.

Indeed, many Brazilian states had already forbidden the use of asbestos. National consumption had dropped to around 125,000 metric tons, and exports began as early as 1980. In 1984, Joseph Milewsky, discoverer and founder of the mine, was half retired. He had given up his job as General Manager and dedicated to a new task: he started to create export markets. The latter really grew from 2,000, in order to compensate the reducing national consumption. In 2011, SAMA happened to export 175,000 metric tons, wildly helped by the closing of Canadian mines. Brazil became thus the third world asbestos producer, just behind Russia with 800,000 tons, China with 400,000, but ahead of Kazakhstan limited to 150,000. In November 2017, the Supreme Court definitely banned asbestos extraction, industrialization and commercialization of asbestos in the entire territory.

According to local authorities, it seems that not any asbestos illness appeared in Cana Brava since the 1980s. For long, the company carried out a very avant-gardist politic regarding prevention, with a high rate of investments due to protect workers. Until the 1960s, the allowed rate of asbestos fibre concentration was high and practically uncontrolled. In the following years, it was progressively reduced to 0.2 fibre/cm^{3}.

After this long parenthesis linked to the big Brazilian asbestos mine, indispensable to so many industries, let us revert to our aim, asbestos-cement.

Around 1964, Brasilit built a new factory in Belem, on a site offered by the State of Pará. Roads were also in very poor conditions and transporting water tanks required important transport means. The situation made people in charge to actively build a harbour on the Tocantins River, nearby the factory and some 40 miles from the ocean. Then, barges loaded with 250- and 500-liters water tanks could navigate to the Amazonian net and even to the Atlantic.

In 1971, Eternit created a factory on Goiania, near Brasilia and in 1972 another one in Colombo PR. At the best of its growth it will totalise a workforce of 2,000.

In 1974, Brasilit opened a new plant in Capivari, near Campinas/SP which along the following years, became a model factory. Charles Biaggi was also a leader in the growth of the plant.

In 1980, Eternit purchased Wagner SA. This one had built a premade drywall factory in Manaus/AM, two years earlier.

From that same year, the" Brazilian Industrial Security" began to export itself towards neighbour countries including to Duralit in Bolivia. This latter received at least twice a year, visits from Brazilian Medic Audits, who performed all kind of required inquiries. It was the beginning of consciousness of dangers in asbestos-cement plants, as well as in every company using asbestos as a raw material[135].

In 1992, the two main Brazilian asbestos-cement manufacturers got married. Brasilit was bringing 55% in the bride's basket and Eternit 45%. The result was: "Eterbras Tecnologia Industrial" who had to manage nine asbestos cement factories and benefited of the experience of two big international groups, Saint-Gobain and Eternit.

During 2001 and up to 2003, Brasilit liberated of SAMA mine and willing to fall into line with Northern countries methods, started to work on new technologies. He was

[135] Information received from a Bolivian engineer who spent most of his career in the fibre-cement industry in different Latin-American countries.

the first one to reinforce flat sheets and water tanks with synthetic yarns (*CRFS: Cimento Reforçado com Fio Sintético*). Soon after he followed the same line with every sheet bound to industrial buildings.
He finally decided to give-up the too expensive PVA and to create his own polypropylene fibre, named "*Brasifil*". In that purpose, he built his own polypropylene fibre manufacture in Jacarei/SP.
From experience, engineers knew that polypropylene generated plenty of trouble due to its lack of mechanical resistance and lack of adhesion to cement. *(The elastic modulus of polypropylene is five times smaller than that of PVA and ten times smaller than that of glass-fibre*). But they found out a yarn surface treatment in phase with the stretching, which strongly improved the polypropylene "frictional adhesion".
Sheets made with cellulose and fibres issued from the such treated yarn, would present a sufficient breaking resistance and a better ability to suffer shocks and flexibility, all qualities highly appreciated when dealing with roofing.
According to some experts, the risks of multiple cracks would appear with a smaller load than with PVA fibres. The future will tell us if the defect, frequent with fibres of low elastic modulus, will fit local and international norms applied to roofing materials. With thermal and mechanical resistance, it will certainly be the same thing that sheets will meet throughout their life.
In May 2007, Eternit S.A. partly withdrew from EterBras-Tc-Industrial Ltda, kept only 20% -Capivari and Contagem/minas Gerais- At the same time, factories of Goiás and Rio Janeiro were transferred from Eterbras-Tc-Industrial Ltda to Eternit S.A. In December of the same year the divorce was complete. Eternit withdrew totally from Eterbras, each partner pulled out. The smart wedding glorified in 1992, barely lasted more than ten years.

In 2007, Saint Gobain, owner of Brasilit, sold his shares in Cana Brava Mine to Eternit. The mine changed its name from "S.A. Mineração de Amianto" to "SAMA S.A. Miradors Associadas". Eternit was then the only fibre-cement maker in Brasil to use asbestos and to export chrysotile from Cana Brava mine to other Latin American, Asian and African countries.
Brasilit, newly independent, specialised in asbestos free products and tried to obtain asbestos ban from the Brazilian government. This gave rise to a quite complete war between former partners.

In 2007-2008, Brasilit began to produce new and diversified products based on "New Technology", with synthetic fibre type polymer and its own polypropylene fibres.

In 2012, on the occasion of his 75th birthday, Brasilit built a new line of corrugated sheets, 5 to 6 mm thick, in Recife/PE. In 2015, SEROPE/RJ, Brasilit created another new line to produce "TOPCONFORT" corrugated sheets, with high thermal qualities, which permit to reduce inner temperature by some 4°C (39F).

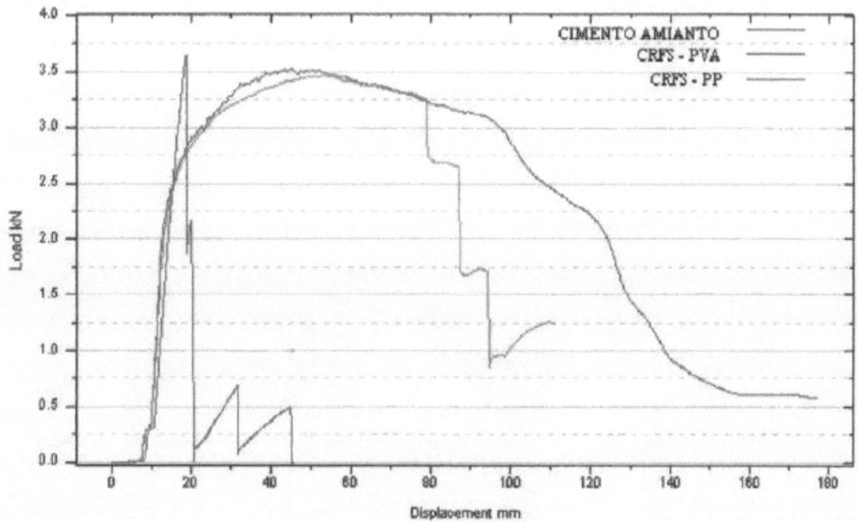

08-007 Load scheme Asbestos-cement/ PP fibres/ PVA. Private archives

A visit on the website of Brasilit will be the best way to find out if they still progressed since those lines were written.

C for

Canada

We already said a lot about Canada in the chapter dedicated to asbestos.

Yet, this country, formerly great asbestos fibre producer, now turned the page and presents pretty buildings made of contemporaneous fibre-cement, free of asbestos. Such is the case with the present villa with its so North-American style.

08-008 Big American style villa. James Hardie Website

Yet, we must add that everything is not quite as rosy in term of quality and it happens that certain lack of stability in size of the new products create big problems between architects and builders or fibre-cement manufacturers.

On the other hand, an article read in "le Soleil" dated August 25th 2012, informed us about the trend of the time and explains how fibre-cement sheets are used to make smart floors in villas or flats. The present sample seems quite significant as well as in aspect of easy laying.

08-009 Contemporary fibre-cement floor. Website "Le soleil »

Let us revert once more to the old days. The British Group Turner Asbestos Inc. which was very present in the mine of Asbestos City, Quebec, had his own sheet factory there. It was dispossessed by the local government in favour of "*Société Nationale de l'Amiante*". Production was maintained till 1988.
In 1976, in Montreal, Atlas Asbestos Inc. gave birth to Atlas Turner Inc. which specialised in roofing sheets, wall sheets and water pipes.

At the same time, in the western part of the country, thanks to security of supply offered by regional asbestos mines, in Mission, British Columbia, appeared Turner Building Products, whose goal was to reduce the risk of US imports.

In 1992, John Zacharias, English aristocrat and former executive manager of Atlas Turner, created Cemfort Inc. in Montreal, at the very moment when North America and Europe where fighting with the asbestos crisis. In the following years, around fifty former Atlas Turner members, supported by external financing, converted the plant to asbestos-free cement production. This made it the only factory that made asbestos-free roofing sheets and siding in all of Canada.

08-010 Atlas Add

Four years later, in 1996, financial problems, added to the fact that the plant was located right in central Montreal, they had no choice but to move. This is how Cemfort had to sell the main part of his reserve and machinery. As a result, production ceased and Cemfort saw his activity limited to sell imported products, among which CELA Boards (*steam-cured panels*) imported from Mexalit, Mexico.

In 1998, again in Montreal, appeared Finex INC. created by Helios Muñoz, Mexican engineer, formerly working at Cemfort and Atlas Turner, but without conflict with his former boss. The newly arrived company worked with the same range of products, imported part of Mexalit and some from Plycem USA. Finex, helped by Mexalit R.&D., also developed new products such as FINEX PANELS. The latter are intended for balconies, were frost and thaw resistant exploiting the snow isolating properties. The product, called "Finex NT Panels" met a wide success in west Canada and is still prosperous by now.

In 2009, a new comer also appears in Montreal: *Matériaux de Bâtiments* ARKEA Inc. specialised in transforming panels. They propose modern final touches and hung products perfectly able to compete with the best European and Japanese similar commodities. They can supply the needs of eastern Canada and northeast USA

08-011,08-012,08-013, FINEX frost resistant products.

08-014 Courtesy of ARKEA..

Finally, in 2015, in the above regions, confidence in fibre-cement products is back. They now fit commercial, residential and even sanitaria's buildings.

Central America

It seems that the history of fibre-cement in Central America began when Eternit from Switzerland created Ricalit in Costa Rica, near San José capital town of the country. Within a few years, this young company became a leader in the area thanks to its light prefabricated houses.

08-021 Prefabricated house in construction

Let us emphasize that in this part of the world, as well as in other developing areas, the type of building, inspired from local conventional architecture, is rather well convenient, being easy to settle at reasonable cost.

It can also be noted that most companies around here build their names with the same "it" suffix already met in so many countries: Ricalit (Costa-Rica), Nicalit (Nicaragua), Hondulit (Honduras), Techolit[136] (Panama)...

Around 1988-1990, Stephan Schmidtheiny who had recently sold his share in European Eternit(s) came over Central America. He decided to give up asbestos and to work exclusively with new technologies. This decision made that Central America can be considered as a pioneer in the art of asbestos less fibre-cement. This is how the name "PLYCEM"[137] progressively gave birth to such specialized plants, not only in Central America, but also in Ecuador, in Venezuela and later-on in Argentina, Colombia, Mexico and even in USA. No doubt that Plycem, supported by Stephan Schmidtheiny was among the first asbestos free fibre-cement producers, if not the first one ... they were not without meeting problems as, flat sheets especially, used

[136] Techo: Spanish for ceiling.

[137] Plycem: it seems that the name was built, somewhat like "plywood", with the three fist letters of cement.

in wall panels for prefabricated houses, were not always defect free, which sometimes generated big disputes between manufacturers and buyers.
In 1999, Plycem merged with Amanco specialized in PVC pipes. The resulting company took the name of Amanco, while the word Plycem remained the name of the asbestos-less fibre-cement. As often happens, profits did not reach the expected result of the merger and a divorce occurred in 2004. Stephan Schmidtheiny then sold Plycem to the Mexican Mexichem, which belonged to Elemantia conglomerate, due to become one of the giants among building products manufacturers in the three Americas[138]. We will revert to this point when talking about Mexico and USA.
When he sold Plycem to the Mexicans, who were producing both asbestos-cement and fibre-cement without asbestos, Stephan Schmidtheiny faithful to his former commitment to sustainable development, made them sign that the former Plycem plants would stay exclusively to asbestos-less products. One year later, Mexicans were planning to make asbestos-cement in the Salvadorian plant. Stephan Schmidtheiny immediately took action to prevent such situation. Thanks to the contract, he won, thus avoiding the return of the damned fibre against which he was fighting for so long.
In August 2015, an article from "*El Pais*" informs that Plycem from Costa-Rica is the first company in the speciality to insert robots in its process, reducing thus its costs by some 16%, which grants it to remain competitive on international markets.
In Panama, I had the opportunity to visit several times the very sympathetic little Techolit, which was producing corrugated sheets and moldings for the local market. Around 2010, it suffered the same destiny as most independent companies and was taken over by the Mexican Elementia. Soon after, the plant ceased activity and was dismantled - machines were broken- in aid of the Colombian plants belonging to the group. Having eliminated a regional competitor, they could improve their sales.
In El Salvador, a small company, founded in 1963 by local investors, probably with technical help from foreigners during the first years, is still competing with the Plycem plant, but using mostly asbestos. According to my information, this company, named Duralita, recently began to replace asbestos by "New Technologies".
In fact, the history of fibre-cement in El Salvador is one of the least clear. One point is certain, the country is very small, yet two companies have been competing for

[138] Three Americas: North, Central and South. Let us remind that the generic name of Latin America, often called L.A., refers to the huge region going from North Mexico to Tierra de Fuego.

decades, one exclusively with asbestos and the other one exclusively with "New Technologies" i.e. without asbestos.

Chile

In 1937, independent investors founded Pizarreño. Like most young companies in the fibre-cement business, Pizarreño began with one sheet machine and one pipe machine.
Few years after, it was taken over by Eternit from Belgium and later-on entered Etex Group S.A.
Pizarreño plaid an important part in the Chilean history, as it actively participated to rebuild after several natural disasters suffered by the country.
In our 21st century, the company also entered the post-asbestos era using new technologies. Supported by the group, it pursues modernisation efforts. At the same time, James Hardie, opens a plant in the country, considering it as a strategic place to penetrate the promising South American market. In spite of an interesting development, the *"Australian construction materials company James Hardie Industries Ltd, has sold its Chilean unit fibre cement to local construction materials company Compamia Industrial El Volcan SA, for US$15.8m. This was reported on July 11, 2005"*[139].
By now, the small companies which I had the opportunity to visit at the end of the 20th century seem to have totally disappeared from the Chilean market.

China

Such as with so many industries, China was long behind occidental countries, but has now widely caught up the gap. Knowing the huge demand for every type of building one can just expect a very large development of the fibre-cement industry.
For long, China was wondering if converting to non-asbestos production was worth. Many researches were held in view to make one's mind. Due to high demand from the building industry the conversion is coming. Chinese industrials now choose steam-cured technics, probably under pressure from Australia or USA.

---

139 Source: Cemnet.com 2005 / 07/11

According to "IIBCC 2006 - Sao Paulo, Brazil. October 15 - 18, 2006. Universiade de Sao Paulo & University of Idaho: Sao Paulo, 2006":

"Due to the sustained and unremitting efforts made by many foresighted and sagacious Chinese scientists and experts, the old aspect of China's fibre cement industry has been changed greatly; which means that the long period of the solely existing fibre cement product, AC, is closed, and a new transition period of AC and non-asbestos fibre cement, co-existing, has appeared. The main no-asbestos fibre cement products being developed are CCA, VRC and GRC. Though at present the share of these non-asbestos products in the annual fibre cement total output is still less than that of AC, the manufacturing processes and application techniques of the former are getting more and more ripe.
Furthermore, many people in the country have already understood that certain properties of non-asbestos fibre cement are even better than those of AC. Undoubtedly, non-asbestos represents the development of the fibre cement industry in China".

Even though difficult to obtain reliable data from the country, a contact with a fibre-cement felt maker, PURUY Felt and Accessory for FC Industry in Guangzhou, helps to get some dependable information about the market. Let us read what the sales manager friendly wrote me in 2017:

"*In 2013, there was a report for this industry: 93 factories, 149 lines*

From 2013 to 2017, I think there have 10 new factories with about 14 lines, so now total about 100-105 factory, with 150 lines, but the running lines is about 90 to 100.

Now, most of them are not use asbestos, just some little small factory use asbestos, I do not have the currency number, even they use asbestos, they are transfer to no asbestos now, this is by the market demand.

At least, my customers, more than 50 factories, only 5 to 8 use asbestos. The rest use original pulp or wasted Kraft pulp.

1: *In China, most of the production is calcium silicate board, and this may 60 to 70% of the market, the rest is fibre cement sheet.*

2: *Most of the production is flat sheet, just little corrugator around 5% I think*

3: most of the production line is flow on (round-cylinder machines) about 80%, only 20% is Hatschek machine.

*4: The total production is about 500 million SQM (1220*2440*6mm) per year".*

After reading the above, one of my friends, great expert in the fibre-cement industry, took pleasure in figuring out the following:

"Knowing that a 6mm thick square meter weighs some 10 kilos, the Chinese yearly output could be estimated to some 5 million tons, that is around five times what France was producing in the most glorious years. It seems a lot, but if we relate it to the population of the two countries, it is not much. Future development is still very wide".

08-022/ 08-023 Making roll and wide sight on China's fibre-cement machines

In the same development line, the German Company WEHRHAHN, already quoted, informs:

"*Tianjin Zhongjing*" from Tianjin received two high capacity Wehrhahn flat sheet production lines for air cured and autoclaved fibre cement boards. The plant has already been installed and started production in 2019. In the plant a newly developed cement which offers many advantages in comparison to standard Portland cement is used." End of 2020 two additional lines with even higher capacities were ordered and will be supplied in 2021.

Colombia

During WWII, Eternit Switzerland set up three factories for asbestos-cement in Colombia, in Bogota, in Cali and in Barranquilla. With such a program covering almost the whole country, Eternit was soon accused of monopoly. In order to clear its name, Eternit set up a new company called Colombit located in Manizalez (Caldas, centre of Colombia). In spite of that, the situation became soon unsustainable. In 1979, Saint-Gobain, who was already well established in South America, but not in Colombia, purchased Colombit.

Some members of Colombit did not appreciate the change of owner. They dissented and created Manilit, a new company, with a small plant located just a few minutes from Colombit, with similar plans and machinery. Financing was supported by "Luckers", group specialised in chocolate and powdered coffee.
The Monopoly game was not yet over.[140] In 1989, Saint-Gobain resold Colombit to Eternit Belgium and newly bought from Eternit the three original plants of Bogota, Cali and Baranquilla. In the next years, Saint-Gobain, reorganised the plants, reducing the number of machines and at the same time, modernising the remaining ones with the most recent technics available in its French factories.
At the turn of the millennium, Saint-Gobain sold the three plants to the Mexican Cemex, king of cement. For how long?
When I write the present lines, the three plants created by Eternit in 1942, sold to Saint-Gobain in 1979, resold to Cemex in 2000, are now in hand of another Mexican, Elementia, whom we will largely talk when dealing with Mexico.

Cuba

We could see earlier that PAM had created Société Perdurit in the Cuba. PAM had then convinced some technicians and foremen from its Mexican factories to migrate and to work under the management of Jean Argouges, who later-on became Technical Manager of Everitube in France.

---

[140] The history of fibre-cement in Colombia was obtained from a friend engineer who spent the main part of his professional life in this industry in Colombia and neighbour countries.

Twenty-five years later, the President of Everitube (recently passed at Saint-Gobain), signed a contract and the corresponding technical specifications, for the supply of a ready for use factory in Castro's Cuba. The contract included one sheet machine and two pressure-pipe machines, one of them up to 600mm (23.6in) diameter and the other one up to 1200 mm (47.24in). Staff training in the factory of Andancette, Drome, France, was planned, as well as industrial commissioning by French specialists.

Machines were entirely designed and produced by the so called *"Département des Travaux Neufs"* (more or less: New Works Department) from Everitube. As usual, the manufacturing of the machines was entrusted to subcontractors in mechanics, boiler making and other specialists. The group already had quite a good practice in the matter, acquired in its own branches, but little in communist countries, except lately in Algeria. Throughout the setting and commissioning, the French detached technicians who had to learn how to think and work under the Castro regime. A succession of unexpected events will brighten or disturb journey, stay and start of the machines... The young Cubans who had been training in a French plant were to take part in the start together with the specialists sent by the management of Everitube. The commissioning appeared more difficult than foreseen, for the simple reason that unlike usual in France and Europe, the asbestos used in Cuba was exclusively from Russian origin. Importing from South Africa was politically unthinkable. The problem was finally solved thanks to South African blue asbestos imported through France. Transit via our country warranted the right political colour.

Due to the urgency, an experienced engineer from the Andancette factory had to be seconded and sent to Cuba in a hurry. The specialist in fibre-cement pipes was invited by the management to reach Cuba on January 1st 1978. Caring not to leave his family behind during New Year's Eve, he obtained to postpone his flight to January 2nd. We were then at a time when flights to Havana, via Madrid, were far from being daily. January 2nd, by chance, was one of the "no fly day". Everitube's people made their upmost to find a solution for the engineer and the foreman who had been designed to accompany him. Two seats were reserved on board a Russian flight coming from Moscow, bound to Havana with a stop in Rabat, Morocco, before flying over the Ocean. To reach Rabat, the two assigned volunteers had to change flights in Casablanca. Surprises began for them in the boarding room at Rabat Airport.

Few people on board, no more than ten or twelve travellers. Boarding was done in a quite military style, rather different from the rush we know today. On the plane, the steward ordered each passenger in a very authoritative manner to take his place. To their surprise, in spite of the few passengers, the two men had their seats assigned at each end of the cabin. After a longer than usual wait and a normal take off, during which no one moved nor spoke, the crew informed that belts could be unfasten. Immediately a joyful noise began. Passengers exchanged seats, got bottles and chess games out of their bags, giving birth to a friendly atmosphere which was going to last most of the flight. Our two French travellers appreciated the opportunity to join seats. After a few hours, the plane began his descent toward the ocean, generating some fears which vanished when it finally landed on the smallest Caribbean Island, I mean: Barbados. The stop did not last more than some two hours, which allowed the local technicians, time to check and maintain one of the engines. The flight could then reach Havana without any trouble. There, the change of temperature, already noted in Barbados, became obvious. Welcome appeared just as formal as boarding in Rabat the previous night. At the police desk, passports were seized and replaced by provisional Cuban ID cards supplied by the" *Departamento de Desarollo Industrial*" (Industrial Development Department). Soon after, foreign currencies -French Francs in the case our travellers- were equally seized and replaced by "convertible *pesos*". Their number was carefully registered in an "exchange-book". The "convertible *pesos*" should allow some purchases in the few hotel shops and access to some restaurants. Cuban "wages" as agreed in the contract were to be paid in local non-convertible *pesos*.

After passing the police and customs desks, the meeting with the French team took already place on site. First of all, however, a visit to the "*Departamento Sanitario*" (Health Department) had to be made to do blood test and to check that foreigners entering the country were not importing any illness that could affect Cuban citizens. Separated for a while from their fellow co-expatriated who had to rush to their jobs, the two newcomers found themselves locked up without any explanation, due to their total ignorance of Cervantes' language. New reason of anxiety for them as they did not understand the reason of their detention. A few hours later, their colleagues came back to the airport to pick them up and were surprised not to meet them. They began to search...and soon found out that this kind of quarantine was due to their transit via Rabat...but why?

Let us remind the reader that at the time, isolated by the US blockade, Cuba could only survive thanks to the "great Russian brother", whom it was totally dependent. Russia was at war in Ethiopia. Russia needed soldiers, preferably with the right skin...consequently Fidel Castro would steadily supply troops to his generous godfather. When soldiers would come back home after their African stay, they supported

a health quarantine. Our two Frenchie's had flown on an African flight between Casablanca and Rabat, their passport wore a Moroccan stamp...The mistake was cleared and the two victims of the Russo-Cuban friendship were soon liberated.
Then came the road to the factory located towards Pinar del Rio, west of Havana, surrounded by banana plantations and a cement plant, one of the very few industries in the country at the time[141]. Workshops are built in tropical and equatorial style, that means without exterior walls, roofs are just sustained by poles where needed. The result is a natural air conditioning. Being given the ubiquity of water in the fibre-cement process and prevailing temperature, mosquitoes were quite happy with the flesh of a northern French young man whose skin was so appreciated by such ambient situation. *(The quasi impossibility to find anti-mosquito cream locally soon became a real nightmare. By luck, journeys back home soon helped to solve the problem.)*
At the end of the day, still in company of their colleagues and carrying their luggage, they could at last reach their first hotel, located in Havana. It is not a castle, but it is reasonably acceptable. Rooms had been booked by *Departamento de Desarollo Industrial*. In the days that followed, professional activity became more organised as our friends began to discover the Cuban way of life. Legendary American cars of the fifties were still almost brand new and amused streets and avenues with their poorly maintained coating. One must admit that Cuban mechanics were making, and are still making, miracles to keep them running, even in our 21st century, for the great pleasure of foreign visitors. Some Ladas and a few Renault, imported by European companies participating to the country's development, boosted the local motorcars range.
Downtown, freedom seemed to be the rule, only a few police officers were to be seen. Yet, everything was quiet. Observing carefully, our friends discovered at each street corner of the old town, the presence of CDRs[142] watching every movement.
Food was free to buy, but locals willing to buy anything else could do it only at work with rationed tickets. Restaurants were classified in three types:

-reserved for locals, *(ordinary people)*

-reserved for technicians *(mostly Russians, Czechs or French)*

-reserved for nomenklatura and foreigners *(owner of the famous exchange book allowing them to pay with convertible pesos)*.

---

[141] Then, main activities in the Island were: tobacco, sugar cane and banana.

[142] CDR: Revolution Defense Committee. In each of these observation points, two or three guards watched, and noted what was going on, 24/24. The State knew everything.

Here comes the time to explain about this strange currency. We noticed that travellers had confiscated their currencies upon arrival and that the amount was recorded in an "exchange book". In order to avoid any kind of trafficking, the possession of foreign currencies was strictly forbidden. If a foreigner who was tied to the system wanted to pay for anything -in a habilitated place- he would pay with the convertible pesos he received, but the person in charge of the location would subtract the amount from the "exchange book". When leaving the country, the foreigner will receive the balance indicated on the exchange book in the currency that he indicated on arrival.

Nomenklatura and our expatriated are welcome in restaurants reserved for customers who can pay with currencies. One can remember the most famous of them, *La Bodeguita del Medio* and *La Florelita,* both dear to Hemingway' heart...and let us add *El Tropicana.* When people from the nomenklatura arrive, foreigners must vacate the best tables immediately. When the restaurant is full, customers, not high class members, may be invited to leave the place.

After about ten days in the Castro paradise, our pipe specialists heard from their hotel staff that they had to gather their things and leave the place. Very surprised by the order, they tried to understand and found that their rooms were only booked for ten days. A new visit to *el Departamento de Desarrollo Industrial* confirmed the situation and granted a new booking in another hotel. There, it appeared at first sight, that it was more a dive than a hotel. Access to rooms, beds with torn sheets, shower-rooms without heads but with holes in the walls that allowed a view of the next room, gave them no desire to leave their luggage in place and even less to sleep there. Quite desperate, our friends did not foresee any solution until the moment when, by pure luck, the manager of the Everitube's *Travaux Neufs* Department appeared. He was a very experienced man in the Cuban way of life [143] and he could manage his own funds. He took our friends to his own hotel, *El Riviera*, one of the few in Havana with norms close to European standards. It was located in *El Malecon*[144] and was a former pre-independence American palace. Unfortunately, it had not received a renovation since 1961. According to the old rule, the big elevators

---

[143] I heard that he had made so many flights between Europe and Cuba on board the old B707 of *Cubana de Aviacion,* that he memorized the numbers of the seats to refuse due to their so poor condition.

[144] Malecon, famous seaside promenade and avenue in Havana, and in many other places in the Caribbean area.

were operated by authorised bellboys. They had their rooms on the 5th floor and decided to visit the surroundings, which led them to discover that all access to the various stairs was blocked, forcing residents to use the elevators. Consequently, no one could leave his floor without being noted by the bellboys in charge... Moreover, they also discovered that each floor was assigned to a group of people being in the country for similar reasons, industrial development, State representatives from a certain country but not from two different ones. The only solution to leave one's room was to transit via the elevator managers. The country's security depended on this wise precautionary principle.

D for

Denmark

You remember Russian Tzar Nicolas II ordered three factories to Denmark in 1910 and that at the beginning of WWI and the revolution of 1917 they had not yet been delivered. One of the pipe machines involved was not lost for everyone.
The Danish Company FLSmith started with the manufacturing of the order that they had booked from their Russian buyer, but the arrival of the war gave them no chance to pass the machines on. They were packed and forgotten for years. Thirteen years later, in 1927, one of them was unearthed. FLSmith then decided to create their own fibre-cement factory: Dansk Eternit Fabrik in Aalborg. The new company achieved a great success and produced millions of square meters of slates for Danish roofs and beyond. Let us point out that the company was among the first to use micro silica (or silica smoke) in fibre-cement products. The Danish market became insufficient, the company developed exports and searched for a new name: CEMBRIT, which they claimed to come from CEMBRI, an ancient Danish tribe that conquered a great part of Europe and even threatened the Roman Empire a few hundred years BC. In our time of 21st century, CEMBRIT has become an industrial empire, with factories in Poland, Czech Republic, Hungary as well as in Finland and with commercial activities in Germany, France, Belgium, Spain, Estonia, Ireland, Latvia, Lithuania, Norway, the Netherlands, Romania, Russia, Slovakia, Sweden, Ukraine and United-Kingdom. The Group is also implanted in USA, through American Fibrecement Corporation.

E for

Egypt

A professional journey to this country around 1980, was the opportunity to visit an asbestos-cement factory, named URA-MISR[145], in the suburbs of Cairo. It left me with some memories.
I was introduced to the technical manager by the local agent who represented then our company in the country. The only theme that seemed to interest both men was the best way to discretely transfer an envelope containing a few US dollars from one pocket to the other...
In the minutes that followed the conversation (*translated from Arab to English by our agent),* which, as far as I know did not generate any business, I obtained the agreement to pay a visit to the workshops so that I could foresee a technical-commercial offer, prelude to any future order. A member of the staff, a foreman I presume, with some knowledge of English, took me inside the factory. There, I discovered a world which seemed nearer to that of the 19th century described by Charles Dickens, than that of the finishing 20th century...
Years later, I read in a newspaper that the plant was closed. Fifty-two former workers who were laid off and asbestos sick, without a pension, expected the help of a NGO to survive for a short while...
During the same journey, I also visited another asbestos-cement factory, which the Spaniard Uralita had just sold ready for use to a local group. Thanks to my friendly contacts with Uralita, our company had obtained orders for the pipe-machine felts. There I met some acquaintances belonging to my Spanish customers and everything was all right. I left the place with confidence for the future.
About 30 years later, while I was inquiring with the idea of writing the present book, I learned that the factory had been closed for several years. The owners begged Uralita for possible technical and financial help in a potential restart of the factory. A Spanish engineer, friend of mine, told me:

---

[145] MISR (pronounce miser) is the former Arab name of Egypt. Many Egyptian companies include the word in their name.

"I was sent there by the management of our company, in order to check the practicability, make a technical report and estimate the possibility of reopening the site. What I saw was nothing but desolation: abandoned and rusted machines, great number of parts evaporated... circulatory drainage, asbestos dust, pieces of fibre-cement, pieces of old felts scattered around. All that made me advise my management not to enter in his "rat trap". By luck, they trusted me and dropped the project. I believe that the factory never restarted".
No comment.

Ecuador

There were few things to say about this country, where I knew two fibre-cement plants, one in Quito and one belonging to Eternit in the harbour of Guayaquil. There too, manufacturers of home furniture were interested in fibre-cement as testifies the herewith advertisement.

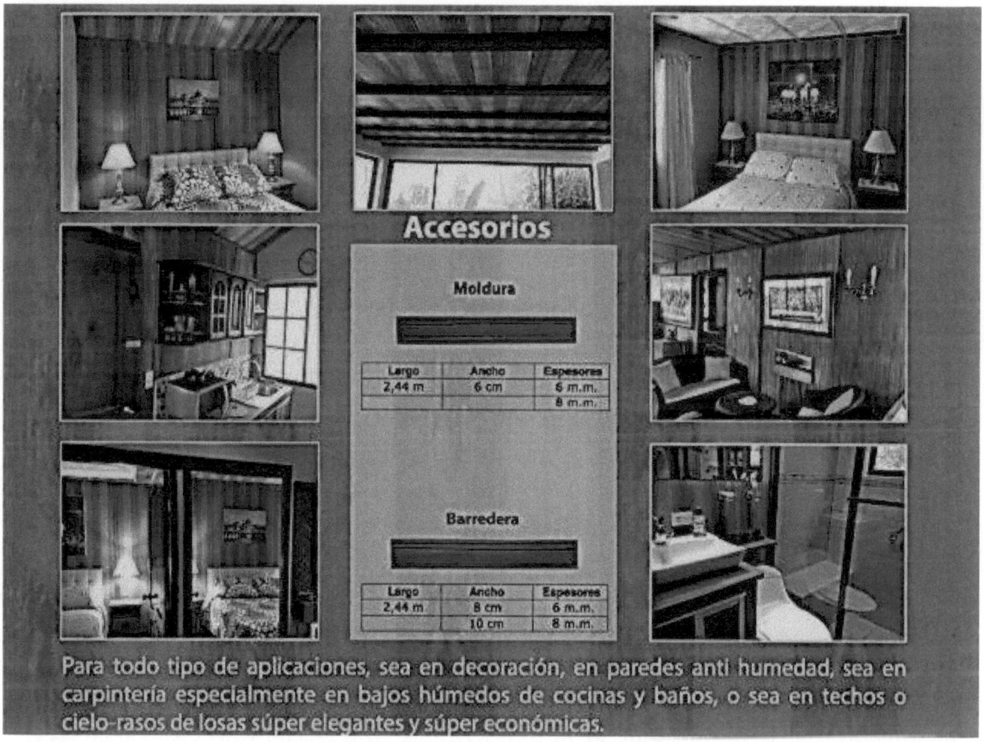

F for

France

I am not going to say much about my country as I have already talked about it a lot. Let us remember that after knowing many asbestos-cement factories and a strongly correlated crisis, only one company nowadays produces fibre-cement, now clearly without asbestos. I mean Etex, Eternit Group, who runs three plants as follows:
-Thiant, recently moved to the nearby Haulchin, near Valenciennes, Hauts de France.
-Saint-Grégoire, suburbs or Rennes, Brittany, specialised in slates.
-Terssac, near Albi, Occitanie, specialised in tiles.
Yet, some other worldwide groups commercialize their products in France. Such is the case of CEMBRIT with offices in Briançon, Elementia via his importer SCB located near Orleans, James Hardie, whose products are sold by several companies that sell building material. Note that James Hardie came up with a clever solution to escape the poor image of fibre-cement in France. James Hardie offers his products as: *"composite -cement"* .
A small detail comes back to my mind. We talked about research carried out in different countries at the end of the 19th century to develop industrial products that would meet the needs of the building industry. We can add that France also had researchers on this subject. In fact, a small factory making "ferrocement" existed in Bonnières, Ile de France at the very beginning of the 20th century. It was not asbestos-cement, but a mortar reinforced with steel structures, most time just a kind of grillage. This product, supposed to have been invented in 1848, met a wide success years later in ships construction, even if one does not expect to meet the so called "cement-boats" on the sea. I personally knew one of them on the river Seine in my youth.
If in another issue the production of sandwich panels in Europe was forgotten for a while, it has now been revived with the new cellulose-cement panels, which cover a structure made of polystyrene (PSE) or polyurethane (PUR). Exterior cladding is supplied in rough or coloured versions, for instance in Glasal. In the field, as in many others, the competition is

very tough, aluminium sheets, enamelled mirrors, laminated panels of all kinds, thin natural or artificial stones fight in the construction market. They are called *EdR* (for reinforced elements) when used outside. Fibre-cement producers entrust the manufacturing of sandwiches to subcontractors, such as branches of ISOSTA Group. Obtained products are all certified by " *Avis Technique du CSTB".* They also receive a quality certificate testifying their quality.

Companies that import their products in France must submit them to SCTB up to twice a year. Controls are carried out at the importers' premises and can last up to 48 hours. Sometimes, pleasant meals may take place during the visits.

In March 2015, the magazine Capital published an article from Philippe Eliakin Calles: *"Pour désamianter la France"* (or "How to quit asbestos from France").The summary is quite clear: "In the name of the Precautionary principle, our country choose the most severe anti-asbestos rules in the world. Consequences might well be catastrophic." Readers interested in the theme will be welcome to read the article on Capital website with help of the above link or Qr code..

G for

Germany

Although Germany has played an important part in this industry, let us return for a few minutes to SAIAC[146] , which was already quoted Chapter I.
In December 2007, Maria Roselli, published "*Die Asbestlüge*"[147] which unveiled the testimony of Nadja Ofsjinnakova about her life, and that of many others, in the forced-labor camp of Eternit during WWII.

Extracts:
"As soon as 1929, manufacturers had created a cartel called SAIAC in order to optimise their industry. Of course, the cartel had its address at Eternit in Niederurnen, and Ernst Schmidheiny, owner of Eternit Switzerland, was the President.
In Switzerland the subject is almost a taboo, quite unknown. But this asbestos cartel participated to the deportation organised by the Nazi regime, and benefited of the Berlin forced-labour camp.
During WWII, the Deutsche Asbestzement AG *(DAZAG) employed more than five hundred people from the Berlin forced-labour camp in its Berliner plant. Half of them were foreigners, French war prisoners and Italian civilians during the first time. From 1942, they were mostly women from Eastern Europe who were working in the Eternit Factory, according to the name appearing on the doors of the camp.*
Nadja Ofsjinnakova was certainly the only survivor of the camp…when Maria Roselli went to her retirement home in Riga, Latvia, and recorded the testimony of the former deported woman".
Fortunately, modern Germany now offers us a much better image. Should I only mention Wehrhahn, source of the following picture and supplier of such machines as well as of complete AAC production plants worldwide.
Forming a fibre cement sheet is shown in the picture, which is part of the machine, behind the workers' legs. One can see, among other things, that the

146 SAIAC : Société Anonyme Internationale de l'Amiante-Ciment : Asbestos-Cement Ltd.
147 Die Asbestlüge : The Asbestos lie

round sieves transfer the moist paste onto the felt, as explained on the first pages of the book

08-027. Modern Wehrhahn Hatschek machine.

Workers on the footbridge help imagine the size. Courtesy of Wehrhahn.

I for

India

The Indian fibre-cement market counts a great number of local enterprises, in very different size. Some of them offer asbestos-cement pipes. Such is the case of:

-ARI Intratech in Jaïpur,
-Chamdran Enterprises in Mumbay
-Danross Industries Private Ltd in Gurgaon
-Capexil in New DELHI
and many others.

Let's take a look at two pages extracted from a company's advertisement whose name is not necessary, which develop arguments that any western association that cares about our health would practice greatly:

> *"Asbestos is stronger than steel and its thermal coefficient matches with cement. It provides a homogenous, microscopic, three-dimensional reinforcement in AC. Thus, it can withstand more than 15 times the water pressure than a concrete pipe of similar thickness.*
> *"Has a durability of about 100 years & gains strength with age !*
> *"The safety factor increases unlike in alternatives where it reduces.*
> *"Lower initial costs and longer life means cost per year is 1/4th to 1/10th of its alternatives.*
> *"Is robust, leading to low storage cost.*
> *"Able to withstand the transverse stress arising out of internal pressure, including water hammer (pressure surge).*
> *"Has high flexural stress i.e. high crushing strength. Therefore, it can be used safely under heavy external loads and heavy traffic. It can withstand back pressures even when the pipe is under vacuum.*
> *"Its low coefficient of expansion helps counteract the expansion and contraction that results from the temperature fluctuations in the water. The unbalanced pressure created by bends and closures including gates.*
> *"It is corrosion free – both chemically and electrolytically.*

"Friction loss is low and high carrying capacity is permanently maintained, as the pipes are not subject to encrustation or tuberculation. C value = 150. As a result, there are significant savings in performance and pumping costs compared to most other pipes.
"It has low thermal conductivity / excellent insulation as a result, water conveyed by these pipes remains cool even when the ambient temperature is high.
"These pipes are comparatively light and easy to handle and transport.
"Jointing is easy & the joints have flexibility (up to 10 degrees in AC couplings) to obtain a satisfactory joint (unlike most alternatives).
"Easy & inexpensive to cut and tap and to use & maintain. (Often known as permanent & maintenance free pipes.)
"Are heat & fire resistant.
"Can withstand chlorine treated drinking water without tuberculation erosion.
"Excellent for intermittent water supply where corrosion, tuberculation & erosion problems are acute.
"Excellent for tropical countries.

08-028 India Asbestos-cement pipes stowage. Web Indiamart

Indonesia

When international enquiry agencies inform that Indonesia will soon be the sixth economical country in the world, there is no surprise that Etex set a second fibre-cement plant in the country. *(Gresik Plant is operative since the beginning of the 1970s)*

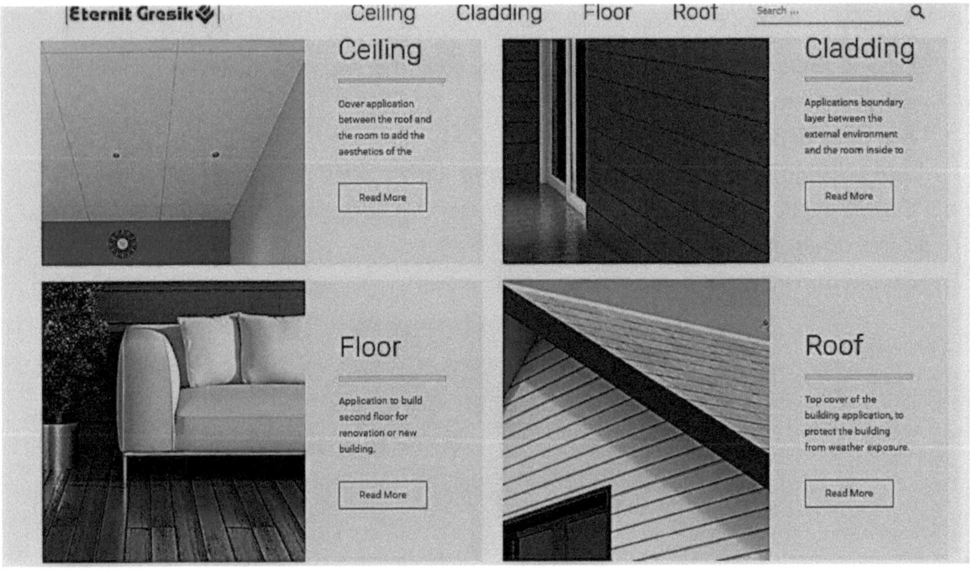

08-029 Eternit Gresik advertising.

To strengthen its position in the construction industry, whose very high potential has been recognised worldwide, the group decided to make a major investment in 2015. In March 2016, Her Excellency Princess Astrid from Belgium, representing the Belgium Economical Commission, a company of Indonesian and Belgium authorities, inaugurated the new plant of Kerawang, near Djakarta.
When we rely on the Eternit Gresik website, it seems that the offering covers all the needs for the house, floor, interior walls and sidings.
According to the same Website, every sign related to the development in this part of the world is green and there is no objective reason not to believe it. In Indonesia, not

only is the population growing, but also a middle class, want to improve their well-being. The expected impact on the building market sounds great. Additionally, the taste for plywood seems to be waning as wood become scarce and prices rise. Fibre-cement takes advantage of the situation. *(Up to the discovery that cement plants are taking a too big part on pollution...)*
One manufacturer presents a video showing the skill of fibre-cement panels that resist flood. Access to the video seems somewhat uncertain. A search on PT Global Indonesia Asia is worth trying.

Iran

Once again, *"Patron de Droit Divin"* by President Roger Martin[148], opens his memories to develop, always with humour, the history of asbestos-cement in Iran. He tells us:
"At the end of the 50s, PAM -which at the time was not yet Saint-Gobain- was not present in the Middle-East. The creation of Iranit in Teheran in 1947, was a good opportunity for the French group to penetrate the industry in this part of the world, so promising in term of needs in pipes and building. The young company received two sheet machines and one 5 metre pipe machine. The president of Iranit was one of the managers of the neighbour cement plant.
Thirty percent of the company belonged to the Pahlavi Foundation, directly cared by the Shah and the Imperial Family. Pont-à-Mousson would ensure technics with just 48 percent in order to preserve the susceptibility of other Iranian shareholders".

[148] Roger Martin was President of Pam (Pont à Mousson) and of Saint-Gobain-Pont-à-Mousson after the merger of the two companies.

During a trip to Iran in 1966, Roger Martin had the opportunity to have some unexpected and incredible meetings with the Shah and his family, which led to an extensive project being carried out in 1974 to create a new factory near Isfahan. The latter was a source of incidents which resulted in a kind of intimacy between the Martin family and the Shah's.

In the autumn of 1974, a young engineer from Everitube *(the branch of PAM in charge of Iranit)* was sent to Iran with the aim of taking care of the maintenance of the Teheran factory. It soon turned out that his technical experience was more useful in assembling and starting the new factory near Isfahan, where he was immediately affected.

When he arrived, he found that nothing was ready. There was no office room available to start work. The pipe machine was not there yet. It arrived from France in the following weeks, in pieces by trucks and not in the right order for an easy installation. Truck trips between Western Europe and Iran were a real adventure in and of themselves, usually lasting three weeks. When climbing the very high mountain passages that made it possible to reach Iran, trucks had to drive very slowly, kids clung to the vehicles and robbed everything that could be dismantled, such as headlights and taillights. To prevent such thefts, experienced drivers brought sweets with them, which they offered if necessary. The trip was so risky that certain drivers hid a gun under the dashboard...

On the worksite, the administrative manager was the only one to speak English. His help was often required when detailed orders had to be given from the French engineer to Iranian workers... On another point, the tea ceremony was respected as a prayer. You have to know that tea is a national tradition in Iran, either in public institutions or in the private sector. When the building has several floors, there is a man on each level, who apparently has nothing to do. Still, he watches every newcomer, whether they are coming from the stairs or from the elevator, if there is one. He remembers the room the visitor enters for some reason. A few minutes later, he in turn enters the room holding a tray with the correct number of cups so that everyone is properly looked after.

But let us go back to the beginning of the new factory. Let us admit that work includes a few issues resulting from both local rules and the climate. One of the issues remains especially strong in the mind of my retired informer. He told me the story as we walked through the streets of Paris on our way to *Place de La Concorde*, where Saint-Gobain was celebrating its 350th birthday.

Any plant making pipes needs an overhead travelling crane. One evening the one ordered from France for the Isfahan plant arrived. The man in charge told the workers: *'We will check that tomorrow morning'*. The next day around twelve o'clock, when the crane was out of its case, one of the workers came and saw the man in charge and said: *'Hey chief, we have a problem, the crane is warped'*. General disbelief led the chief to visit him immediately. The notoriety of the manufacturer left no doubt about the quality of its products, but at first sight there was no doubt. Measuring appeared indispensable. A cable was stretched between both ends and a 7 cm (0,4in) difference was noted in the middle. Everyone wondered, scratched one's head and suggested assumptions. A new measurement confirmed the first one. A phone call to the supplier made him think of a joke. Given the persistence of the buyer, after a few days he decided to send a technician to the site. He could only determine the reality of the claim and he too scratched his head. He asked his own manufacturing manager, who in turn swore that the crane was perfectly straight when it was delivered. Then the carrier was asked if there was any possible incident during the road transport. Nothing happened on the way. Finally, it was decided to send the crane back to the supplier.

Our young engineer who was responsible for commissioning the factory was still wondering. One morning, around seven o'clock, he went back to the crane and was surprised that it was straight...The mystery got worse! After thinking about it again, he got it. The temperature rose between the end of the often very cold night and midday in the sun. The distention was different between the girder in the sun and the other who stayed in the shade. The next night no more extension and everything was back in order...

Long after the Islamic revolution chased the Shah and the French from Iran in January 1979, my interlocutor reminds another source of surprise: *"At the factory in Teheran,*

decanting water was sold to sprinkle the nearby cotton plants. The latter were in fairly good health".
I personally remember that the felt company I worked for was very happy with the large orders that came in from Iranit twice a year.

Italy

Italy was one of the first countries where the asbestos cement industry was growing at the beginning of the 20th century. In 1907 the Mazza family built their first factory in Casale Monferrato, Piedmont. As we could see, Adolfo Mazza was a pioneer when he developed the first pressure pipe machine as early as 1912. In the same year a second plant was built in Cavagnolo, Turin.[149]
The arrival of industry in Piedmont was seen locally as a major progress in this poor rural area where misery, starvation and exhausting work were the norm. The location was perfect as the region was well endowed with clay, which was needed to produce cement. In addition, the biggest asbestos mine from Europe, Balangero, was no more than one hundred kilometres (some 62 miles) away.
The inhabitants conceived the factory as a wonderful opportunity to improve their quality of life, thanks to wages offered and work condition appearing less painful than in clay mines, although the working conditions of the 19th century with our eyes of the 21st century are more likely to result in thinking of slaves.
Likewise, new items obtained with asbestos-cement seemed like wonder and progress: roof sheets, poultry shacks, cool-boxes, prefabricated cabins..." anything coming from Eternit was good". Kids were using wastes from the factory to build tree-houses, bags full of dust were proposed for multiple use...
Fabrizio Meni wrote about that time:
"Society was divided into two main categories: a well-educated elite, which held the power, and a mass of ignorant and subdued people who could only count on their own strength and long hours of hard work, which combined meant exhaustion. The elite could not only read and write they also spoke the national language. In contrast, the mass of manual

[149] Most information on this page comes from a press article by Massimo Gigliotti published in January 2013 in the Italian magazine Science Natural.

workers was for the most part illiterate and common bestial dialects not recognised outside their local area: viewed by the elite as work animals with no language, but expletives, whispers or curse words. Today it doesn't seem believable that such a separation was justified by unquestioned anthropological beliefs. The elite assumed intellectual and spiritual superiority since the weary masses were regarded as being inherently stupid with no perception of a spiritual life. But along with the introduction of paid work in industry, came a sense of wealth for the workers. The working-class felt like a bearer of values and for a while thought it was contributing with its own work to the construction of a better world for the future".[150]

During WWI, asbestos-cement products were used in various types of shelters because of their high fire and weather resistance. Finding asbestos fibres, however, was not the easiest and encouraged the search of substitutes. The company built a saw mill and a firebrick workshop in order not to dismiss the workers. After the end of the war, asbestos-cement production was resumed and reached its historical all-time high as early as 1919.

In 1920, Ernst Schmidtheiny, became president of the board. The crisis in the same year was not a problem for Eternit. Only asbestos-cement roofs resisted a very large fire, which offered the company excellent advertisement.

From 1928, pipe production experienced an extraordinary development that lasted until the 1970s and plaid an important role in modernising the network of drinking water as well as sewage.

In 1933, corrugated sheets became the roofing norm for schools, hospitals, gyms and movie theatres. The great success of asbestos-cement products at the 1939 International Exhibit in Munich, Germany prompted the Italian government to classify asbestos-cement as a strategic material.

During WWII, more or less for the same reasons as during WWI, the company searched for substitutive activities to preserve its workers.

Corrugated sheets were very popular in France and nearby countries. Italy, on the other hand, did not develop a high production until the late 1940s. Then the architects did their best and Italian corrugated sheets gained market shares in France thanks to lower manufacturing costs.

---

[150] Fabrizio Meni is a writer and historian who has been working very hard on the Torino Asbestos Trial. What he wrote in the above lines could probably be applied to the history of many countries. For that reason, we will not revert to the point along our "Countries Overview". A visit to his website might be an edifying glimpse.

In 1952 the Mazza family sold their business. Eternit Italy then joined the Franco-Belgian Eternit and partially Amianthus AG, which was also one of the shareholders of Eternit Switzerland before becoming the main shareholder in the 1970s...

In 1975, Stephan Schmidtheiny, the fourth in the name, joined the board of Eternit Italy and began to inspire new technical options regarding sheet manufacturing, especially wet mixing process, which is considered to be less dangerous to the health of workers than the dry process. Later-on, when he sold his Swiss factories to his brother Thomas, he kept his foreign plants, including the Italian ones.

In 1986, in view of the dangers foreseen by the shareholders, it was decided to declare Eternit Casale Monferrato bankrupt. Not long after it was closed and abandoned, along with its waste.

"The idea of "closing the factory" was seen as a crazy plan that would challenge thousands; it would have meant cancelling the "American dream at home" that Eternit represented. To take action, the union had to overcome both the distrust of workers who didn't want to risk their jobs and the hostility of the local people who did not want to support a commitment against the main economic resource of the area. However, this attitude changed when people started dying systematically, even men and women who had never had anything to do with the factory".[151]

Don't the above lines remind us of the closure of the Canadian mine in asbestos or the wedding picture at the lake of the Russian one?

In 1992, Italy in his turn banned the use of asbestos *(for the record, Switzerland did so in 1975)*, and that same year trials began against companies who had used asbestos. Eternit was on the front line.

In Casale Monferrato, cradle of asbestos-cement in Italy, several local managers were convicted.

In 1994, the production of fibre-cement without asbestos, which is known as "sustainable fibre-cement", began. Conceived by Stephan Schmidtheiny, with the hope of maintaining the feature of the previous product.

---

151 Fabrizio Meni: "Eternit in Italy. The asbestos trial in Italy".

On February 13 2012, the Torino Court informed about his verdict on the Asbestos Trial during which had been counted some 2100 casualties and 800 sick people. Stephan Schmidheiny and the Belgium Louis de Cartier were sentenced to 16 years in jail each.

Here are a few more details about the trial, related by Business Human Right Watch on his English website:

"On February 13th, 2012 the court found the defendants guilty of negligence and sentenced them to 16 years imprisonment. In June 2013 an Italian appeals court increased the sentence against Stephan Schmidheiny to 18 years in prison for causing the death of 3000 people. The court dropped charges against the second defendant, Jean-Louis Marie Ghislaine de Cartier, because he died in May 2013. On November19th, 2014 the Italian Supreme Court overturned the lower court's decision and acquitted Stephan Schmidheiny on the grounds that the statute of limitations had expired...

In July 2015, an Italian tribunal referred the case to the Constitutional Court to examine whether former CEO Stephan Schmidheiny could be charged with voluntary manslaughter. On July 21st, 2016 the Constitutional Court ruled that he could not be tried for deaths that had already been the subject of other trials...

On November29th, 2016, charges against Schmidheiny were changed from manslaughter to involuntary manslaughter. On January 12th, 2017, Turin prosecutors appealed to the Cassation Court against a judge's decision to downgrade charges to involuntary manslaughter. On May 23rd, 2019, a Turin tribunal sentenced Schmidheiny to four years in prison...

In January 2019, an investigating judge from Naples started a new lawsuit against M. Schmidheiny. The hearing started in April 2019. On May 23rd, 2019, the Turin court sentenced former CEO, Stephan Schmidheiny to four years in prison for one of the asbestos-related death cases in Italy. He was charged for manslaughter..."

In spite of the above, by luck, the new asbestos-less fibre-cement seems to be prosperous in Italy as can be seen on the following web-site:

Sil-Italiana Lastre Edilfibro

Ivory Cost

Around 1995, a French engineer with experience in asbestos-cement decided to set up a factory in this African-French speaking country. He purchased a downgrade flow-on machine, and assorted accessories, and moved it to Abidjan, Ivory-Cost, Africa.[152]

More than twenty years later, the company now: "*Ivoirienne de Fibre-Ciment*", IFC, seems to be doing well and offers interesting data on its website as can be seen downwards

08-028 Pictures from the Website of Ivoirienne de Fibro-Ciment

The website also provides information:

«*Fibre-cement advantages*:

"-it protects against heat and moisture; it is fire-resistant and also protects against noise.

-IFC is the first enterprise to specialised in fibre-cement in Ivory-Coast.

-Sheets are cut in squares 40 x 40 cm and rectangles 60 x 30 cm for roofs and ceilings.

152 Personal memories of the writer at the time he was invited to supply the relevant felts and visited the Abidjan factory a while later with the aim to get orders for the next felts.

J for

Japan

It is quite difficult to gather information about the empire of the rising sun on our topic. Fortunately, in 1986 two French specialists and good observers visited their Nippon homologues on behalf of the French Everitube. More than thirty years after the journey, one of them made a great effort to memorize and report to us, their most noticeable observations.
First, let us emphasize the pretty name of one of the Japanese enterprises which they met: *Asahi Sekimen*, which could be translated into Rising Sun Asbestos...
One must keep in mind that meetings between professionals took place quite exclusively in dedicated meeting rooms. Visits to workshops were strictly limited to an absolute minimum. However, Japanese companies certainly emphasize the cleanliness and full recycling of waste. In fact, such care was already in the 80s, quality and steadiness had to be one hundred percent according to properly established norms. Conditioning and packaging were treated with the same care. At that time we were very far from the European way. *(Personally, I had the opportunity to note the difference when, in 2016, I visited a French site that has been closed for more than twenty years).*
Japanese production plants are quite small compared to European standards and even larger compared to US ones. No dump around and no waste getting out. The use of asbestos is also strictly observed. One notices that plants are landlocked in urban fabric for most of the time. Knowing the few agricultural areas available in Japan and the high density of population, it is easy to imagine that people looked after their vital surroundings long before we did.
Our visitors also noticed that, contrary to Europe, productivity did not seem to be the main topic. When in France the most modern sheet machines had to produce 300 metric tons daily, or the guy in charge had to report to his management, in Japan respecting norms seemed much more important than the number of tons getting out of the machine.
For this reason, our French visitors did not understand why Asano Slate Corporation near Osaka needed a six round sieves sheet machine, especially since, in the mind of European engineers, production was not limited by the Hatcheck part of the machine

where single layers take shape, but by cinematic problems met downstream, such as corrugating cycle or start and stop inertia of moving parts. They finally concluded that the interest of a 6 vat machine was in thinner monolayers, offering better mechanical properties and less risk of delaminating *(for a sheet thickness equal to that obtained on 3 or 4 vat machines)*. Besides, one may recon that such a machine is surely more expensive to operate, would it just be for the length of the felt and the number of vats to maintain and replace. Still very attentive, the visitors noticed the presence on the machine of a belt corrugating device RCM, quite an infrequent tool, *(scheme chapter III)*.
When they visited the laboratory of Asahi Glass near Tokyo, they noticed the existence of a rather basic lab Hatschek machine. The purpose of this machine was to test special glass fibres being studied with a view to asbestos-free future production.
In Kubota near Kyoto a new surprise; they discovered a “drying process” machine. The first and only one they ever saw in their entire life. It seemed conform to what they knew about thanks to J.P. Guerber’s Asbestos Dictionary published around 1960.They wondered if Mister Kubota couldn’t be a follower of Johns Manville, who is well known to have used this type of machine?
This “drying-process” solution produced small coloured elements in various shapes, which were used in façade embellishing. Such coloration and shape would have been much more difficult with a Hatcheck machine. The visitors were not invited to check the following manufacturing steps, which came just before steam-curing. They left the place convinced that such a technique could not be adapted to other fibres than asbestos, except possibly for glass-fibre.
Their last visit was for Nippon Eternit, one of the three pipe-makers of the country, with Chichibu Cement and Kubota. Nippon Eternit was founded in 1931 by Tokyo Gas Company, then called Japan Eternit Pipe Company, which had acquired a small part of the rights from Eternit Italy. The new company had meanwhile gone its own way and Nippon Eternit no longer had any connection to Eternit Europe. The visit was brief and almost without interest as it did not show anything unknown to the guests. The company disappeared from the pipes manufacturer’s scene in 1997.
At the end it appeared that the main asbestos-cement market in Japan, except for pipes, was for private housing in form of roofing and all kind of façade embellishments. Agricultural and industrial buildings were far from the prominent place that they had in many European countries.
Foreign competition seemed to be totally unknown in the Japanese fibre-cement market. Indeed, where could it have come from at the time? China and even Korea

were not what they now became. Other Southeast Asian countries such as Vietnam, Malaysia and Thailand are a bit far away and not yet highly developed in terms of industry. Moreover, many middlemen such as wholesalers, semi-wholesalers and retail networks were part of the trading circles in Japan, which was usually no advantage for a foreign company to enter the Japanese market.

In the year of the visit, most Japanese asbestos-cement manufacturers were still convinced that *"less than 5% asbestos was not dangerous!"*

No pulp treatment system escaped the keen eyes of the French visitors. Though the country was already a big producer of PVA and PE *(polyethylene),* asbestos-cement manufacturers did not seem to be ready to replace asbestos by anything. Stream-curing was not much frequent which made them still farer from possible solutions such as developed by Cape Boards in UK or James Hardie in Australia. In 2004, the asbestos ban in Japan let the bells toll.

When they left, our visitors had in mind that Japanese manufacturers were only interested in environment and quality, but did not feel like entering in new technologies, like Australians, or FBK *(Faserbetonwerk Kolbermoor)* who was making the unbreakable *"Wellcrete"* in Bavaria, Germany. They seemed to be sitting comfortably in their inner market, which they envisioned as sustainable. But who would have foreseen in France in 1986 that asbestos-cement would not ahve more than ten years ahead to live?

NICHIHA

A visit to Nichiha website reminds us that Japan banned amphiboles *(blue and amosite)* in 1995, but the company completely abolished their use back in 1981...

The company gained an international reputation when it established Nichiha Inc. USA in 1998 and a production factory in Macron[153], Georgia in 2007. Its aims are not limited to USA alone, as they are now represented in Canada, Mexico and Russia. What a change since the visit of my fellow citizens thirty years ago!

[153] No common point with the 2017 elected French President, except name.

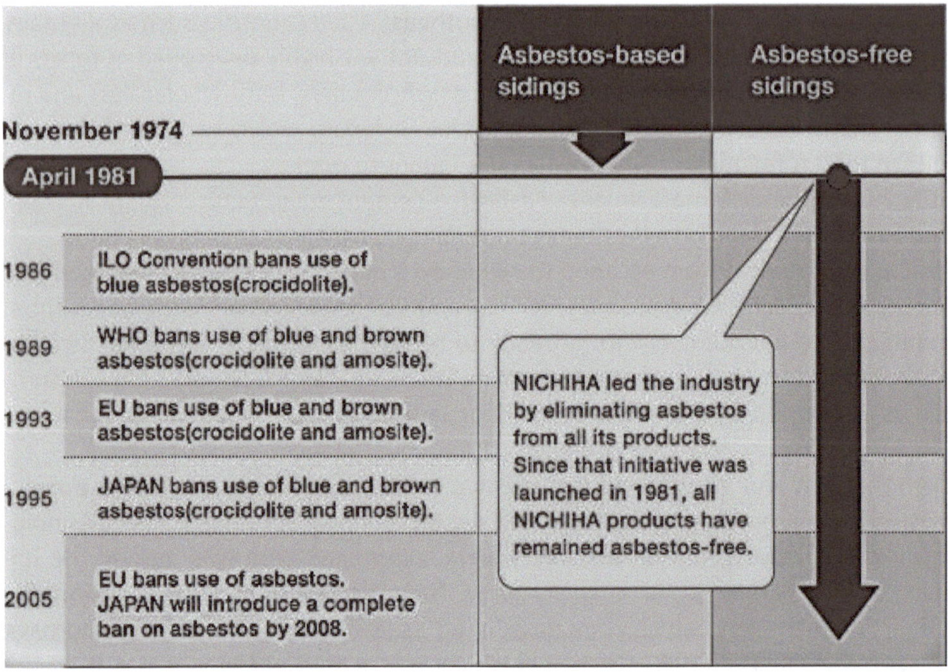
Asbestos-based sidings
Asbestos-free sidings
November 1974
April 1981
1986
ILO Convention bans use of blue asbestos(crocidolite).
1989
WHO bans use of blue and brown asbestos(crocidolite and amosite).
1993
EU bans use of blue and brown asbestos(crocidolite and amosite).
1995
JAPAN bans use of blue and brown asbestos(crocidolite and amosite).
2005
EU bans use of asbestos. JAPAN will introduce a complete ban on asbestos by 2008.
NICHIHA led the industry by eliminating asbestos from all its products. Since that initiative was launched in 1981, all NICHIHA products have remained asbestos-free.

08-029-08-030 Coloured buildings. Nichiha Website

L for

Luxemburg

As far as we know, Luxemburg never had any asbestos-cement plant on his soil, though many buildings were roofed with artificial slates.
The truth is that the Grand Duchy prefers to accommodate a reserved organism in charge of managing the wealth of a large international group, unfortunately without exclusiveness, as other states are highly competitive in the same field.

M for

Malaysia

With 300,000 square kilometres and a population of 31 million, Malaysia provides accommodation to two recently settled fibre-cement manufacturers.
Hume Cemboard Industries Sdn Bhd, "HCI" for short, belongs to the Hong Leong Industries Berhad group, "HLI" for short.
I won't go into the history of Hume Cemboard Industries, which changed its company name six times between 1980 and 2014. As I write this book in 2019, it seems that the Company name HCI still exists. They say they are deeply committed to the sustainable principles of their mother company HLI.
The company is proud to be the first fibre-cement producer to obtain the MS ISO/IEC 17025/2005 according to the ILAC Malaysian standard and to have only commercialized products with the PRIMA ™ ISO/IEC designation. I assume that you, as an attentive reader, are wondering what PRIMA™ is all about. Well, it's quite simple:

PRÎMAdrywall™ Overall, Prima Drywall consists of fibre-cement products from PRÎMA Fibre Cement Building

Solutions, which are intended to build an entire house without mortar. The system includes external walls, floors, ceilings, interior panels and even kitchens and bathrooms.
Another fibre-cement manufacturer, LIUX, was proud to offer Reinforced Fibre Cement (CFRC) with multiple advantages and sold in many places like Europe, the Middle East and others, but it seems to have disappeared now. Perhaps it has been taken by a more powerful group.

Mexico

In 1930 a rich Spaniard founded a company almost in the centre of Mexico City with the assistance of friends of the Spanish[154] asbestos-cement field. A few years after the factory moved to the west of the town, next to a cement-plant. Then manufacturing started for corrugated sheets, *tinacos* -water tanks- and small pipes used as down gutters, using the air-curing and watering process.
In 1936, Eureka became "*Techo Eterno Eureka*" and, with the help of Lazaro Cardenas, President of Mexico, received the first credit from "*Banco Urbano y Obras Publicas*", which gave it the opportunity to produce numerous pipes required for new infrastructures, especially in port towns.
Eureka later set up more factories in Guadalajara and Monterey, the country's largest cities.

Asbestos de Mexico

In 1946, Johns-Manville, main manufacturer of asbestos-cement in USA, founded "Asbestos de Mexico", with the participation of Mexican contractors and bankers. So a newcomer was born near a large cement factory in Barrientos, a northwest suburb of Mexico-City. Johns-Manville dropped his steam-curing technique, which he used in every manufacturing process, to the basket. Along with the growth of the country, the company created new factories in Guadalajara, Hermosillo and Villa

154 Although I have not received the proof, I have good reasons to believe that the "Spanish friends" came from Uralita, as it is near Eternit: Why wouldn't Eureka have been partially associated with Eternit?

Hermosa under the "Asbestolit" brand. Years later, the company adopted the name "Versalite" with the idea of forgetting asbestos.

Asbestos de Hidalgo

In 1979, Rafael Gutierrez, a former asbestos-cement distributor supported by local bankers but without any help from anyone in the business, founded "Asbestos de Hidalgo" in Tizayuka, Hidalgo. He built a small semi-manual factory and began selling his products under the "Asha Gumont" brand.
In 1987, less than ten years after their birth, when they had to relocate the plant from Mexico City, Eureka purchased Asbestos de Hidalgo, turned it into Eureka, deleted the brand, modernized the factory and let it grow.
In 1993 the Eternit Group purchased Eureka, developed Etex image, but kept the Eureka brand. It retrieved a large pipe machine from one of its factories in another country *(it could be the pipe machine from the French Caronte factory near Marseille, which was already at standstill ...)*
In 1997, Eternit purchased asbestos-free sheets and moulded products *(Cela, Siding, Cemplank)* both for the Mexican market and for exports.

Mexalit

In 1952, PAM,*(French group Pont-à-Mousson)* founded Mexalit. The man in charge was Adrian Fouillet, a French assigned citizen, who collaborated with a Mexican-Texan investor. They purchased a very small factory, which they soon closed, built a new-one in Santa-Clara north of Mexico City, and entrusted the management to the first Mexican engineer, Jaime Olivera Zubizarreta. The newly arrived company also took advantage of the growth and needs of the Mexican market to develop and modernize itself with the help of the French Everitube.
In 1964, it opened new production facilities, for example in Chihuahua, north of the Mexican Republic, where a pipe plant was created, managed by a young French en-

gineer Alain de Metz, whom we talked about in the prologue. Roger Martin, President of Pont-à-Mousson participated to the official opening and recalls a memorable feast to blow-out[155].

Mexalit developed the steam-curing technology for pipes from the beginning. The fact of mastering the process *(inherited from Johns-Manville)* was indeed of a great help when the time of entry came in 1985/86, first in corrugated sheets and later in flat sheets without asbestos, as well as in plank-siding and others. It also helped break into the increasingly demanding US market.

In 1986, supported by the French Saint-Gobain *(in the meantime PAM and Saint Gobain had merged)* had his main shareholder, Mexalit decided to create a commercial branch in the US, Maxitile Inc. in charge of selling corrugated sheets and CELA slates (CELA for *Celulosa-Auto-Clavada*, meaning: steam-cured cellulose), finalized by Mexalit with the help of the French pilot plant.

But not all within the group was gold. A retired technician who had travelled a lot between the French and foreign branches of Saint-Gobain-Pont-à-Mousson told me at his home near Paris:

"In 1987 in Mexico I had to work on pre-cutting the corners of small corrugated sheets[156] . I was allowed three weeks to finish the job that previous colleagues had worked on for a long time without success. When I was practically finished within the deadline, I was invited to pack the machine in wooden cases. I inquired why such an urgent task had to be packed as quickly as possible. The answer was: 'Eternit does not do it anymore, Mexalit does not do it anymore either!".

Internal disputes arose while searching the markets in the United States that same year and competing with James Hardie's flat sheets. Mexalit R&D had to work at full speed to develop a Hatschek machine making roll from a pipe made of asbestos-cement with a large diameter. The technicians formed a structure on the "pipe making roll" that imitates wood. The invention gave birth to "Maxiplank Sidings" and the right to sell them in the United States. Thanks to the system, Mexalit was the first company to offer "wood-like fibre-cement siding" on Hatschek machines for the Mexican and American markets.

155 "Patron de droit divin », page 216.

156 The idea was to face the claims made by U.S. retailers and roofing manufacturers who were dissatisfied with the need to cut sheet corners on site.

08-031 From left to right: Fernando de Aragon, director of Mexalit R&D, Jean Louis Befa president of Saint Gobain, Alain de Metz president of Mexalit and General Delegate of Saint Gobain for Latin America in front of the "wood looking like, making roll".

Mexalit is moving towards new technologies:

From the 1960/1970s, Mexalit managers knew, like most managers of companies that used asbestos, that sooner or later they would have to stop using asbestos and find other solutions. But which[157]? They also knew that whatever solution, it would be more expensive and perhaps less efficient.

---

[157] See chapter V « The Crisis ».

To develop the market and ensure the transition to new asbestos-free technologies and prevent Eureka *(at the time Eternit Mexico)* from taking over Versalite, Mexalit bought it in 1988. The next year Mexalit closed the Hermosillo and Guadalajara plants, only kept one pipe machine in Barrientos and one sheet machine in Villahermosa and strengthened the market department. A short time later, Mexalit eliminated the pipe machine and deleted the name of the brand.
Saint-Gobain held several large fibre-cement companies in Latin America. In 1998 two of them joined their efforts and set up an R&D project called BRA-MEX, which also included collaboration with some European fibre-cement producers, with the aim of finalizing a PVA or PE fibre that could replace asbestos in the short term.
In 2003 their research resulted in the birth of Brasifil in Jacarei, SP, Brasil *(already quoted in pages dedicated to Brasil).* The plant had a capacity up to 9,000 metric tons per year of the new polypropylene fibre, sufficient to produce 500,000 metric tons of NT-FC *(New-Tech Fibre-Cement).*
At the same time, Mexalit R&D signed, independently of BRA-MEX but with the same goal, a technical agreement with a Mexican company from Merida, Yucatan, which deals with textile polypropylene fibres and micro fibres for reinforced concrete.
A while later they obtained a new microfibre called "Fortaleza Fibre". The annual production of the new microfibre could reach more than 4,000 metric tons, sufficient to manufacture more than 250,000 metric tons of NT-FC. The new fibre fulfilled the high requirements for low elongation and strong adhesion to cement, which are required by "NT-FC".
Let us revert to a few years earlier. In 1996, ten years after founding of Maxitile Inc. in Los Angeles, California, USA, the needs of the market and the competition from James Hardie, the Australian market leader in NT-FC, Mexalit had been induced to plan a specialized plant in Ciudad-Juarez, Chihuahua, very close to the US border, strategically located next to a cement plant and major silica-banks.
At the same time, the US CertainTeed, other Saint-Gobain branch, was thinking of a similar project, but wanted it to be built in the US. After long discussions between them, the two members of Saint-Gobain asked for arbitration against their parent company. CertainTeed won, the plant was built in White-City, Oregon, USA , and inaugurated in October 1999. Mexalit was among the guests[158]. All of this happened

[158] Most of the above historical details belong to a Mexican engineer who lived the story from 1967 to the present day.

before the big changes that would take place within the great American fibre-cement groups.

"America para los Americanos"[159]

In 2001 the Mexican Eternit plants were transferred to Mexalit. The name Eureka was retained, but Eternit's name was deleted. Since the turn of the millennium, companies have not stopped changing hands. In the first year of the 21st century, Mexalit and Eternit Colombia passed from Saint-Gobain to Antonio del Valle, a powerful businessman, banker and prominent Mexican who was already one of the shareholders of Mexalit. He knew the fibre-cement world and was also an important shareholder in Bital Bank. Duly advised by Alain de Metz, he became the exclusive boss of Mexalit and consequently of Eureka and of Eternit Colombia. The following year he brought the three companies together in his own holding: Kaluz.

2002 the banking age: Antonio del Valle suffered a hostile takeover against Bital Bank and sold his shares. Soon after, Bital was retrieved by the famous giant HSBC. In 2003, Antonio del Valle, who worked as a banker, founded a new bank, "Banco BX+", thereby strengthening his Kaluz holding and at the same time his focus on fibre-cement and a few other activities.

From 2004 to 2008 Mexalit and Eternit Colombia, which had been in the hands of Kaluz since 2000, were brought together. Mexalit purchased Duralit from Bolivia, Techolit from Panama and Plycem from Honduras, Costa-Rica and El Salvador as well as Eternit from Ecuador.

In 2006, Mexalit finally set up a Maxitile plant in New-Laredo, Mexico, right on the eastern border with the USA, for the export of Maxiplank Siding.

In 2009, a new holding, ELEMANTIA, appeared, founded by Antonio del Valle and Carlos Slim *(the most well-known Mexican according to Forbes).* Elementia dedicates its activity to gradually gather important companies of Latin America in the fibre-cement and construction field. The goal can be achieved pretty soon.

In 2014 Elementia acquired Allura, USA, founded in 1986, and also three Ex-CertainTeed fibre-cement factories in Oregon, North Carolina and Indiana.

---

[159] It was disappointing to hear that the exclamation of the Mexalit's general manager and Saint- Gobain delegate for Latin America missed the promising project that had required so much effort from his Mexican team to the benefit of their US colleagues.

The new asbestos-free step: Ex-executives announced that it has been planned for several years to completely eliminate asbestos by 2017[160] so that not any group of plants can manufacture and market asbestos-cement products.
New niche markets: The earthquake in September 2017 that hit Mexico City and high risk areas such as Chiapas and Oaxaca gave me the opportunity to discover the new life of the aforementioned SIPS fibre-cement sandwiches and EPS boards *(EPS: Expandable PolyStyrene)*. Let us spend a few minutes to revert to such boards.
In 1986, Mexalit, always with an eye on the attractive US market, inquired about the old European technology which consisted in jamming a block of expended polystyrene, or polyurethane foam, between two fibre-cement flat sheets and obtaining isolating sandwiches that could be used as a non-metallic structure. The system was known as "SIP", for "Styreno Isolated Panel". Boards can be obtained in various thicknesses according to the level of phonic or thermal isolation and structural ability desired. Sizes result from width and length of available fibre-cement sheets. The most frequent are:
1.22 x 2.44 m (4.00 x 6.56ft)
1.22 x 3.05 m (4.00 x 9.84ft)
1.22 x 3.66 m (4.00 x 10.57ft)
We notice that the 3 available lengths fit the 3 frequent heights of floors.
Other boards made of different materials were tested: plaster, wood, wood-cement, even magnesium, but finally fibre-cement proved to be the best for this type of product.
During part of the decade of 1990, Mexalit exported SIPs made of CELA FC through Maxitile Inc. from Miami, Florida for various uses, but transportation costs were an obstacle to the activity. The US branch finally decided to deliver plain flat sheets of CELA PC to American SIPs makers. Nowadays, Allura and James Hardie supply the US market with such CELA FC boards.
Now we can revert to the sad event of the earthquake of September 2017, when I discovered the existence of a young company named Duratherm, which is located in Puerto-Escondido, Oaxaca, a seismic area near the epicentre and in which houses are built with this type of sandwich boards that resisted much better to the shakes than neighbouring traditional houses.

160 The author has good personal information, I think the goal was achieved in time.

An experienced and implied researcher, who has dealt first with asbestos-cement and later with new tech fibre-cement since the 1980s, provided me with extracts from quality production and the use of Duratherm. Glad to reprint it there as tests performed on SIPs by the famous University of Berkeley, CA, USA, confirm their ability to withstand seism up to 10.5 on Richter's scale.

08-032 Duratherm Video by Berkeley university

08-033 Construction of Duratherm buildings

SIP's SYSTEM

1. One must dispose Fibre-cement boards and EPS blocks

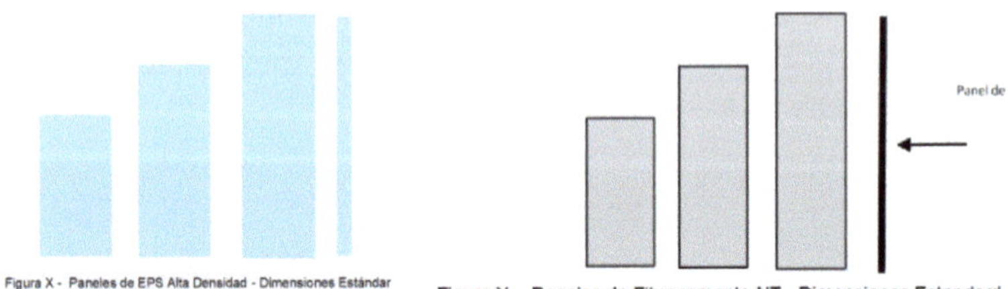

Figura X - Paneles de EPS Alta Densidad - Dimensiones Estándar

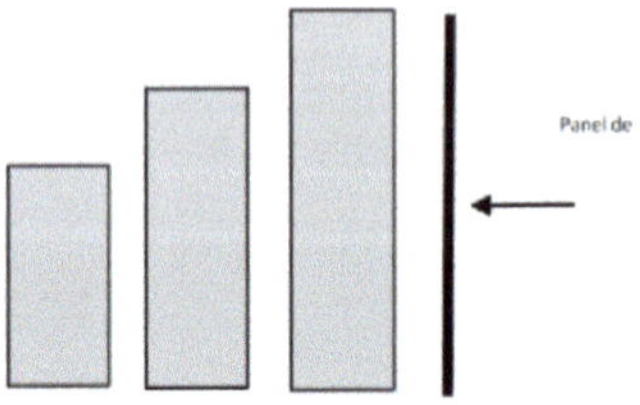

Figura X - Paneles de Fibrocemento NT - Dimensiones Estandard

2 One creates a project, specially conceived for SIP's, with a detailed list of codified materials and building instructions for wood structure or metal profile.

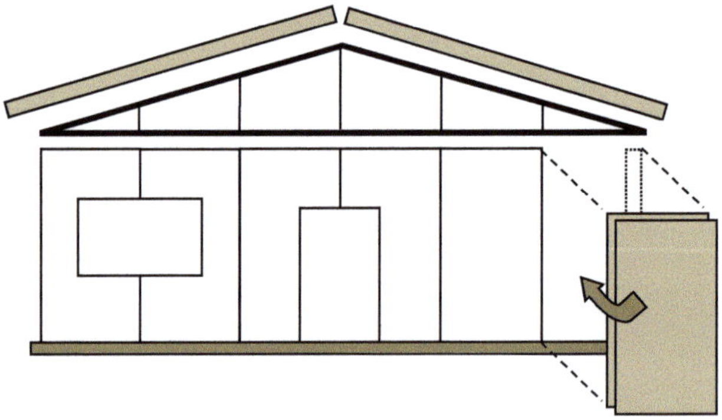

3 Sips' are produced in the plant and encoded.

- One receives an order for each project.
- One secures the process and every required material.
- One prepares certified FC (CELA o NT) and EPS and cuts them at the required dimensions.
- One forwards FC boards and EPS blocks to the manufacturing site.
- Both faces of EPS are impregnated with special adhesive.
- Impregnated EPS is lowered on a FC board and the other FC board covers the opposite side; all are tightly adjusted.
- The just made and encoded SIP is delivered to a multiple press.

- The SIP is now pressed with determined pressure and time, then controlled and stored.

Diagrama de flujo - Producción de Paneles "Durathermsips MR"

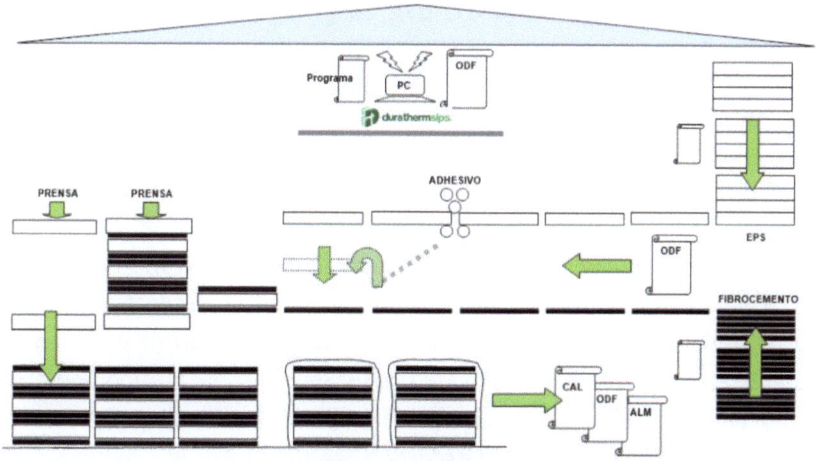

Figura 2 – Diagrama de Flujo

4 Prepared SIPs are forwarded to the work site for building. Once the concrete base is settled, a family house is built in just 3 weeks' time.

5 Isolated SIPs sandwiches for structures are certified for:

Resistance to simple compression, resistance to light lateral load, fire resistance, resistance to impacts, resistance to uniformly distributed loads, resistance to flexibility. Thermic value resistance R and acoustic value IAA. Reflecting "Cool Roof" finishing touch receive IRS certification.

Note: Such development of SIPs technologies was driven by Mexalit R&D since the 1990s and received several "Building Housing Awards".

AFIC *"Association de Industriales de Fibrocemento »* (Fibre-cement Industries Association) including the main fibre-cement companies from Mexico and other Latin American countries are preparing norms regarding seism resistance of SIPs as a material for private housing, in coordination with Universities and Polytechnic of Mexico.

Duratherm Building Systems is the main private houses developer using SIPs technology, with Mexalit FC boards, for up to 3 floor buildings in Mexico.
We should add that resistance of SIPs increases with their thickness, that is, with the thickness of the fibre-cement boards and EPS blocks used. The advised density of EPS - heart of the SIP sandwich - is 0.026 or 26kg/m^3 which actually affects conductivity, thermic resistance, acoustic and structural ability.

December 2020 Publication:

Duratherm building systems The Echo-Friendly Building Specialist

"In 2017,when tropical storm Lidia hit parts of Mexico in the Baja California Peninsula, it destroyed almost everything in its path with merciless fury. Lidia blew away houses, followed by severe flash flooding. While the storm wreaked havoc, a few homes in Baja California managed to escape its rage. The homeowners, who had built houses with Duratherm Building Systems' structural insulated panels (SIPs), found that their buildings remained unscathed. As the storm left behind destruction and debris, these homeowners shovelled the mud out of their house and pressure washed the floors and walls inside. What caught their attention was that the Duratherm panels showed no signs of damage or cracks, despite the winds rampaging all over the area.
Figures from UN report that about1.8 million people in Mexico were homeless or unable to afford a home. "The need of the hour was a high-quality building system that was sustainable as well as affordable for people from different economic groups We did not want to overwhelm people financially with technology that is too expensive," says Bryant.
To this end, the company has partnered with Elementia/Mexalit to produce its fibre-cement board. As one of Duratherm key ingredients, the fibre-cement board, developed by Eng. Fernando Gonzalez[161], is impervious to water and is fireproof, along with a significant number of other advantages and does not require huge amounts

161 Engineer Fernando Gonzalez participated intensely to the pages dedicated to Mexico was of great help to the writer of the present book. We even signed the Spanish version from our two names.

of potable water in the production process. Bryant points out that their board is a practical solution, which works from both an ecological as well as a safety standpoint.

Stronger, Sturdier, Long-Lasting and Echo Friendly

Elaborating on a success story, Bryant highlights a San Antonio based businessman's experience with
Duratherm Building System The businessman built a fishing cabin on one of the little islands just north of Padre Island in Texas. In July 2020, when the eye of Hurricane Hanna went right over his island, the fishing cabin stayed intact while his neighbours' houses were completely destroyed. Despite six to eight foot waves from the hurricane's rage, their panels showed no signs of failure or cracks.
Moreover homeowners can observe a 50-70 percent reduction in heating and air conditioning costs, which positively impacts the carbon footprint as well. Duratherm Building Systems' solutions not only meet the needs of end-users but also offer benefits to builders, developers, and bankers alike. The company s also collaborating with firms that can help lower prices to mitigate the housing crisis in the U.S. and Mexico. Duratherm Building Systems 'major focus lies in offering sustainable housing to those in the low and medium-income categories while continuing to service the upper echelon. The company s also collaborating with firms that can help lower prices to mitigate the housing crisis in the U.S. and Mexico. Duratherm Building Systems 'major focus lies in offering sustainable housing to those in the low and medium-income categories while continuing to service the upper echelon. Due to its zero failure record in earthquake and hurricane-prone zones and cost effectiveness, the team has done projects in the African, the Middle-East and is set to embark on a new one in India. The company is currently working with Eco Sips by S3D, in California, to increase its distribution and assist with opening their franchise in India. Other projects in progress include developments in California, Washington, Florida and Texas, tiny homes and cabins for tourist rentals options, Habitat for Humanity projects and orphanages in Mexico, to name a few".

Plank siding in Europe: I almost forgot to tell you that Mexalit, through his Californian branch Maxitile Inc. is exporting Maxiplank *(Plank Siding)* to Europe. In France, the importer is responsible for very careful painting, and smart façades are obtained with such boards. I could see some of them in southwest France, but there are sure to be more of them in the country.

08-034 and 08-035 Mexalit boards used in South West France

As a conclusion on fibre-cement in Mexico, it is clear that the country will not escape the global tendency or influence of large industrial groups, new materials and the reduction in the number of factories.

Indeed, after having counted at a time, up to four companies and twelve factories, only three brands and four plants subsist in Mexico. Experts say that it is widely sufficient to respond to the present needs and even ensure a constant increase in the near future.

Let me revert to the case of the new Elementia holding company and summarize its ability to bring together famous and recognized fibre-cement companies from Latin America. The following list sounds impressive, doesn't it?

-Eureka Mexico, founded 1932 (or 1930) acquired 2001

-Eternit Colombia, Bogota, Cali, Barranquilla, founded 1942/44, acquired 2000 (via Mexalit).

-Mexalit Mexico, founded 1952, acquired 2000

-Eternit Ecuatoriana, Ecuador, founded 1957, acquired 2004

-Techolit, Panama, founded 1957, acquired 2007

-Duralit, Bolivia, founded 1977, acquired 2006

-Plycem Costa-Rica founded 1964, *(including the factories of El Salvador and Honduras)* acquired 2007

-Maxitile, California USA, founded 1986 by Mexalit, acquired 2009

-Allura USA, founded 1996, acquired 2014 *(including the three plants ex CertainTeed formally Saint-Gobain)*

-And a few more in progress when preparing the book.

Morocco

In 1947, Eternit had founded DIMATIT, an asbestos-cement branch in Casablanca. The company was producing corrugated sheets for roofing and siding, free-standing sheets, special sheets for covering external walls for office buildings, agricultural buildings and formworks. Fibre-cement pipes with diameters of 60 to 400 mm (2.36in to 15.75in) are also manufactured for domestic sewage and rain water. Some of them are used as columns in architecture. *(We also saw the same use of pipes in a 1930s house in Angoulême, France).*

The factory that finally belonged to "Ynna Group Chaâbi" now seems to have disappeared.

In 1982 a new company, "FIBROCIMENT", based in Kenitra, appeared, specializing in sheets, water tanks and window boxes.

From 1996 onwards, influenced by the anti-asbestos wave, such products lost their fame. Activity was reduced as the market was no longer buoyant.

Starting in 2006, FIBROCIMENT gradually abandoned asbestos and focused on new-technologies, especially sheets for sale on the domestic market as well as abroad. This is how it obtained ISO 9001 and ISO 14001 certifications.

N for

Nigeria,

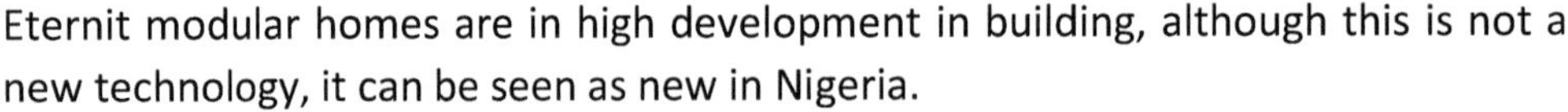

Knowing who is doing what in this big African country revealed quite difficult. The country, which has been independent since 1960, is the largest and most populous country on the continent with 924,000 km² and 186 million inhabitants.
Unfortunately, its recent history is scattered with coups and murders of politicians. It did not prevent the fibre-cement industry to grow, under the guidance of the Belgium giant ETEX. It might be the right time to check out the website and get an idea of the importance of the group.
Until a recent past, Nigeria had four companies: Nigerite, Eternit, Giwarite and Emerite. By now, Nigerite (Etex Group) seems to be the only one still active.
The country draws our attention to the quality of the data that manufacturers offer on fibre-cement prefabricated houses.
We particularly pointed this out in 2017:
"Everything you need to know about modular houses technology":
Eternit modular homes are in high development in building, although this is not a new technology, it can be seen as new in Nigeria.
Modular houses incorporate metal frames in accordance with local standards and use precise and efficient technology specially designed for lightweight construction. This technology uses fibre-cement sandwich panels for main walls and plain fibre-cement for partition walls. Roofs are made from fibre-cement corrugated sheets. The components also include some wooden poles with a rigid structure. The house arrives on site in kit form and can be assembled on a lightweight basis in a matter of days by non-specialized workers without the need for special tools or machinery. Modular houses also support harsh climates without damage and seem to be the best solution in desert, wind, very hot and wet areas.

The advantages of the type of construction also include: the low professional skills needed by the workers, the rapidity of building and, more surprisingly, the high level of burglary protection.

Here is one of the many plans offered. Check out the above website for more plans.

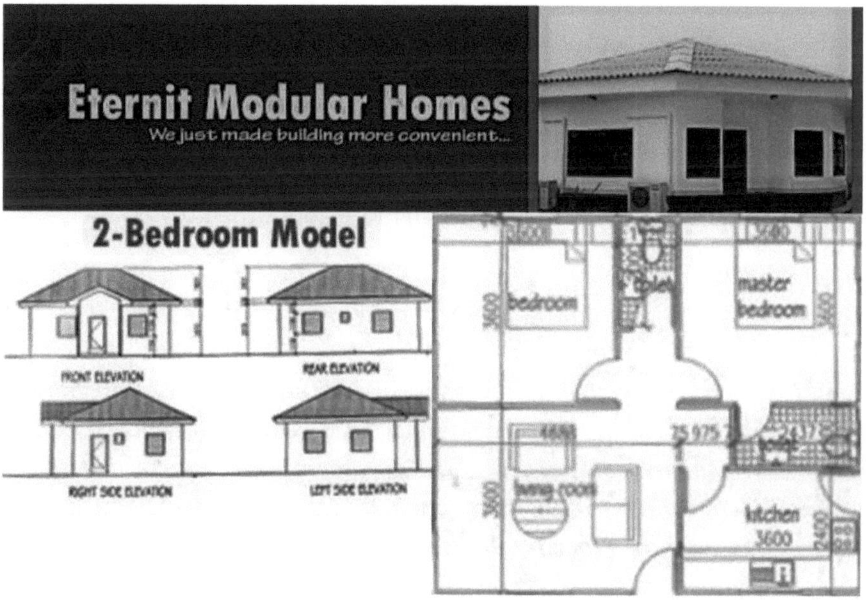

08-038. 2 Bedroom Modular house. Website Eternit Nigeria.

The Netherlands

The fact that part of the country is below sea level does not prevent rare and smart projects to flourish. Attika Architekten in Amsterdam offers real proof.

Here is what to read on the page above, and I hope owners are not worried that I am offering my readers to learn more about the project.

"Drift in Lelystad" consists of eight floating dwellings, for eight families in Lelystad, the Netherlands. After living on the water in their childhood, these families always dreamed of living on the water again.
The families formed a collective partnership called "Drift in Lelystad" (Float in Lelystad) and commissioned Attika Architekten to design eight different but matching floating homes. The municipality of Lelystad, a new town in a polder 4.8 metres below sea level, provided a water location by widening an existing ditch (polders loot).
As all families had their own specific requirements and wishes, each dwelling got its own characteristic size, colour and shape. Direct contact with the water was the key factor in every design: unobstructed views, split level, abundant daylight, water reflections on walls and ceilings, water terraces on different levels and direct access to the water.
The façade panels compose of matching colours. The palette of colours let the dwellings blend into the natural surroundings. All dwellings have two basic colours, thus providing coherence. Colour accents, chosen by the owners, make one dwelling stand out from the other, giving the homes their individual character.
The floating homes were built in Urk, 40 kilometres away from Lelystad. They were constructed in timber frame on concrete caissons and towed over water to their destination. The size of the narrowest lock determined the width of the houses: 6.9 meters".

When I visited the construction site for the first time, I think in 2016, it was specified that façades should be made of Equitone Natura from Eternit. The information now appears to have been deleted...

08-039. Floating houses with façades in Equitone Natura from Eternit, by Attika Architekten in Amsterdam. Attika Architekten Website.

P for

Peru

Eternit came to Peru in 1940 and set up a factory in Cercado, LIMA, where the company owns its main plant in the country. Corrugated sheets, decorative sheets named:" *Teja Andina*" (Andean tiles) and products made from materials other than fibre-cement are manufactured there, which are therefore no part of our topic. After a merger with Gyplac, a new factory was opened in 2014, which, among other things, produces drywall for drywall construction, using a different technology.

Eternit Peruana caught our attention with a rare export method worth telling. The idea originally came from US Department of Energy project called Solar Decathlon. The purpose was to create low energy consumption, a sustainable, inexpensive and easy to maintain dwellings for students. Eternit Peruana was involved in 2015. The project involved 16 se-

lected teams from 16 countries who intended to design a prototype of a cheap apartment that would require a few natural resources as possible, produce as little waste

as possible throughout its life and would be well suited for Caribbean and Latin American families.
Unfortunately, this intelligent project is not yet successful, although a meeting was planned in Cali Colombia in 2019.

Philippines

Although Eternit appears to have been present in the Philippines as early as 1950, I have not been able to find any evidence of its presence in the country these days.
Most great fibre-cement manufacturers of the world are exporting to the Philippine market. Among them, James Hardie seems to be the only one to produce there in a factory located in Cabuyao Plant .
James Hardie offers a very wide range of products, all asbestos free, that cover construction needs from floor to ceiling, partition walls, as well as sidings for kitchens and bathrooms.

R for

Russia

Two Australian fibre-cement specialists, John Cottier[162] and Lieven Alderweireld,[163] explain us in an article about "Safe fibre-cement", published in ISO Focus Magazine of December 2005[164]:

"The basic manufacturing processes for the production of asbestos-cement flat sheet products were developed in Russia in 1896, and three years later an improvement on this process was made in Austria by Ludwick Hatschek".

Such statement, if we exclude the well-known pretention of the former Soviets to invent everything before anyone, could create doubts on the fatherhood of Ludwig Hatcheck in asbestos-cement invention. Whatsoever, it is a great opportunity to take a look at fibre-cement in Russia.

We already saw that Russia is a big supplier of asbestos and that the famous Ural Mountain chain extensively participated to the baptismal name of many companies in Spain, France and UK.

We remember that at the beginning of the 20th century, Tsar Nicolas II ordered three asbestos-cement factories from a Danish company, which could not be delivered due to the international events at the time.

We also know that Russia owns one of the biggest asbestos-mine in the world, which is still in operation. We can add that it plays a major role in the Ourala asbestos conglomerate, which is responsible for the exploitation, and provides employment for a large part of the population of the Asbest City, Ural, Russia.

162 John Cottier, technical manager of R&D at James Hardie Fibre Cement.

163 Lieven Alderweireld, chairman, ISO/TC 77.

164 For easy access to the article, click the link, type "December 2005 Safe fibre-cement" in the search magnifier icon and when it opens, go to page 15.

We can fulfil the present lines, without much risk of error, saying that Russia is going on with high quantities of products in asbestos-cement, mainly pipes both for domestic and export markets, as is confirmed by Nicolas Badiotal in the French daily newspaper La Croix, dated April 25th 2017.

If in doubt, you can briefly visit the website of a big Russian supplier to the building industry who tells us as the following under an anonymous pen:

JSC LATO – fibre-cement. 2013-09-23. Author: administration

"LATO JSC, one of the largest manufactures of building and construction materials in the Russian Federation, produces roofing and construction flat slates, chrysotile pressure and non-pressure pipes (diameter - 100-500 mm, length-4 and5m) and fibrocement boards LATONIT.
Fibre-cement is used to manufacture of wall panels, partitions and cover plates (sheets) which have been used extensively in modern construction for many years.
In 2007, the company "LATO" mastered the manufacture of fibre-cement boards LATONIT. Fibre-cement boards are made of cement, cellulose fibres and mineral fillers, this mixture provides high strength material. The company "LATO" is one of the leading Russian manufacturers of high-quality fibre-cement-based materials".

If Danish technology has failed to penetrate Russia, Finnish technology has done it, with great success if we trust the following:

In 1974, in the town of LAHTI, was founded LTM Company. Soon it became popular thanks to high quality and reliability façade products. In the three decades that followed, LTM's situation changed from a small company to a large one, exporting to a great number of countries, with main markets in Europa, Middle-East, but essentially Russia, where the brand was registered in 1990. Soon it had its sales networks in the country and in 2003 opened its own Moscow office, which had become indispensable.

In April 2006, it was decided to put the most modern coating line in any colour with increased resistance to fire and moisture near Moscow, according to the company.

08-041 Flow-on sheet machine at LTM Moscow- LTM Website

In October 2008, LTM became a holding company and opened a unique factory in Obninsk, near Kaluga and not far from Moscow, with the purpose of producing fibre-cement boards for internal and external use called CEMBOARD, AQUA, VENTUS, etc.

Let's take one last look at the LTM website and conclude:
"The factory uses flow-on[165] technology with the latest European technology for high density fibre-cement boards.
The interaction of the two plants in the same region has reduced the production time of fibre-cement panels and significantly improved logistics. Existence of its own resource base allows us to be independent of imports, which gives the ability to produce high-quality material at stable prices in the shortest possible time".

Such is the paper that we find at LTM... Isn't almost as moving as an address of a people's representative in the old time of the Soviet Union?

165"Flow-on": Remember, liquid paste gets directly onto the felt, via a stock delivery box above and not via round rotative sieves that are located under the felt.

Finland is not the only country supplying current machines to Russia, however.
In 2019, Kazbek from Chechnya ordered a plant for the combined production of flat and corrugated sheets. The flat sheets are autoclaved, while corrugated sheets are produced using air-curing technology. Wehrhahn supplies a finishing plant together with the sheet production line to enhance the fibre cement sheet quality which will generate higher sales prices for Kazbek.

S...for

South-Africa

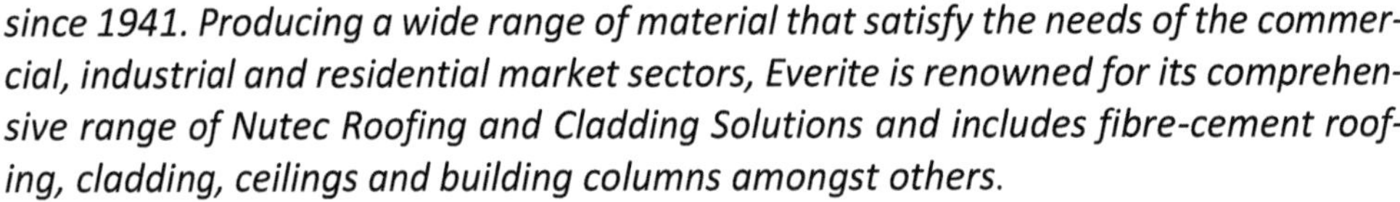

South Africa is tightly linked to asbestos. From the end of the 19th century, the country was exporting asbestos to UK as early as 1883, initially for the textile industry, during WWI for gas breathing apparatus, and finally for asbestos-cement and other industries as we could see pages.
If the country expects numerous casualties due to asbestos, many are due to mine blasts. Spouses of miners and their children should have been most at risk, as fibres enter in houses on father's hair and clothes when they return from work. Few statistics are available. A high level of early death due to tuberculosis occurred among miners. In fact, it prevents knowing how many should have died later on of asbestos reasons.
But in our early 21st century, NT-FC (New-Tech Fibre-Cement), enjoys a bright future, which Everite [166] duly assimilated, as one can read in the first pages of its website.
Quote: *"Everite Building Products, wholly owned by JSE listed Group Five, has been associated with the South African building industry since 1941. Producing a wide range of material that satisfy the needs of the commercial, industrial and residential market sectors, Everite is renowned for its comprehensive range of Nutec Roofing and Cladding Solutions and includes fibre-cement roofing, cladding, ceilings and building columns amongst others.*
Nutec fibre-cement high performance properties and added benefits include: the use of safe renewable fibres; considerable tensile strength with enhanced dynamic load bearing properties; excellent thermal properties; water and wind resistance; hail resistance; fire resistance and resistance to fungus, rodents and acid.
A programme of quality assurance in accordance with the requirements of the International Standards Organisation (ISO 9001:2008)".

[166] Recall that Everite South Africa has no connection with the French Everite that we talked several times.

Spain

We have already written several lines about Spain, but the country cannot be neglected as it plaid a huge role in the fibre-cement industry in the 20th century.
In the 1960/1980, the country counted up to nine factories that made sheets, pipes and various products of asbestos-cement. I personally knew 9 pipe machines, two of which were producing 6 metre (19.68ft) long pipes with a diameter of up to 1200mm (47.25in). A large part of the pipes was intended for pressurised water.
Companies that had originally different funds gradually came into the hands of Uralita, that means into the hands of the March family and *Banco de Santander*.
The factories were located through the entire country as follows:

- Cerdanyola del Valle, Barcelona, mother town of the industry in Spain, sheets, pipes, moldings (*and sometimes art, see chapter VI*)
- Sitges, Cataluña, sheets
- San Vincente del Raspeig, Comunidad Valenciana, sheets and pipes
- Sevilla[167], Andalucia, sheets, pipes and moldings
- Alcázar de San Juan, Castilla la Mancha, pipes.....
- Toledo, Castilla la Mancha, pipes
- Valdemoro, Madrid, sheets, pipes and moldings
- Getafe, Madrid, sheets, pipes and moldings
- Valladolid, Castilla Leon, sheets and pipes.

Note that the factory in San Vincente del Raspeig, commonly called "de Alicante", generated a working class neighbourhood named "*el Tubo*", for "Pipe"! The name remains some thirty years after the closure of the factory.
The Alcazar factory counted only one pipe machine, but what a pipe machine!
With the large and wide pipes it produced, it contributed greatly to the water red tones of the country. They were gigantic pools in which a powerful over-head crane would drop or extract dozens of pipes that matured in water.
I never saw any other factory dotted with such an impressive system.
The first pipe was produced in February 1976 and the last one in April 2002.

[167] During the years the plant was in Seville, it gave me the opportunity to leave Madrid in the early cold morning and land 50 minutes later, as if it were Andalusian spring weather. The charm of taxi drove through Maria-Louisa Park before I reached the factory.

08-042 Uralita, Alcázar de San Juan Factory. The last pipe produced April 22nd 2002. Courtesy of Manuel Rubio, who managed the Plant during years and had the grief of closing it.

The surface of asbestos-cement sheets used in Spain is estimated at 1,500sqm, which is more or less that of the island of Grand Canaria. The total length of pipes buried in Spanish soil is estimated to be 370,000 km or more or less the distance from the earth to the moon.

The factory of Valladolid deserves some specific pages, not for technical reason, but for the first years of its life that are very representative of the wars that competing companies around the world wage between them. Roger Martin tells us with lots of details and a lot of humor in his book, "Patron de droit divin", which has already been cited several times. I cannot resist the pleasure of reproducing the most striking lines, as well for the level of fights violence as for the huge fortune of the March family, owner of Uralita.

Quoted: " 'Pont-à-Mousson' *met a variety of fortunes in the business of asbestos-cement. Only one ended in form of rout, but after all, Napoleon also knew defeats in Spain.*
The idea of an asbestos cement factory in Spain was hanging for long ...market enquiries had been held, though basic were they, they did not urge much, but the project finally took shape. Pont à Mousson had found itself associated with Banco de Santander in Iberit and in Funditubo, a mediocre cast-iron pipe maker, where it had a minority shareholding.
In October 1964, I met the managers of Uralita in Madrid and I assured them that Iberit intended to make its mark in Spain, respecting the rules of a regular competition. It seemed that my words had not blasted enthusiasm between my counterparts. The start of the Valladolid factory which had been quite expensive because the builders had foreseen future growth sparked a severe counterattack. Transporting asbestos-cement is not interesting due to its weight and low cost. Uralita had several factories and a first-class commercial network in Spain, only to then break the prices without much damage in a certain area easily accessible for Iberit. They did it cheerfully.[168].
In April 1966, I decided to discover Spain more attentively and" ...after a large trip around the country..." *I began to learn more about Carlos Caril. At first sight, he was a pleasant gentleman and no problem seemed to bother him. We talked extensively about Iberit and left, convinced that there was no solution outside of the negotiations. I asked him to organise an encounter with the March family on the upper level. A first meeting took place in Paris, in February 1967...When we left, I was convinced that our contacts wanted to get Pont-à-Mousson to give-up asbestos-cement in Spain and they could afford it. A new meeting with the most important members of the family took place in Paris in May..."* It resulted in an invitation for a few days in their Majorca estate. On August 3rd we were welcomed with my wife and several executives from Pont-à-Mousson..." *at Majorca airport, where two Mercedes 600s expected us. The estate of the March family was about three thousand hectares (some 7,400 acres), located in the very south of the island, right in front of Cabrera. The house was built on a small hill. It surrounded an ancient Moorish tower and made me think of a Florentine palace. Dressed up guards, gun over the shoulder, were in line on the side of the terrace where we got out of the cars, facing a portal protected by*

---

[168] I can confirm the words of Roger Martin as at the time, I had friendly contacts with the management of Iberit who was suffering the attack.

two white marble lions. The sons were waiting for us on the threshold, the hostess welcomed us at the middle of the monumental stairs and the householder on the stage of the reception rooms. The ballet was perfectly settled…
…the next morning, Juanito, one of the sons, suggested leaving the "younger ones at work" and that I accompany him around the estate. Our wives and staff would join us at the beach around two. I felt somewhat offended for not being any longer among "the young ones" but I did not really feel like working. Watering possibilities limited the acreage of the estate to three of four hectares -750 or 900 acres- the Land-Rover trip through woods and lands, with moments on a charming and desert seaside created terror among hares and partridges. We reached the beach long before the others. A sailing boat and a motorboat with sailors on board were waiting for us. Juanito, in bathing suit under his bathrobe, ordered the door of a pavilion to be opened so that I could change. The pavilion would have been a perfect place for my family holidays.
…The afternoon and the following day were spent bathing, walking, and idle talks. I was even invited to shoot some hares. Evenings were of the sweetest. Dinner was served on a terrace, cooled by a light breeze, from where we could see the lights of the bustling town far away. Juanito explained that he had negotiated with two or three famous restaurants in Paris that they would each have his main cook on an internship. The result was surprising. One evening, as Renée (the wife of Roger Martin) *exclaimed about the strange taste of a sorbet, the householder disclosed that it had been made from wild strawberries. Very early in the morning, a small aircraft had flown for them to Val d'Aran*[169]*, where regular pickers were operating.*
I should regret that such a memory, perhaps embellished by time, makes the reader think of aggressive luxury. Every detail was cared with an extreme simplicity, without which there is no possible perfection, and corresponded to a kind of work of art, testimony to a way of life long past".

The Hostess *"…let us see some improvements they had made to the estate and some entertainment options they could suggest to their guests: an arena where* novilladas[170] could take place; *an Olympic pigeon shooting with the corresponding dovecotes; a horse stable, next to a quarry, which had been removed after a serious acci-*

---

[169] Val d'Aran is a valley in Catalan Pyrenees near the French border. I feel entitled to wonder if Juanito did not invent the story just to mislead his guest.
[170] Novillada: bullfight in which three-year-old bulls are fought by novice matadors.

dent of one of her daughters, had to undergo long and painful plastic surgery. Spaying water was scarce and the place had been turned into a cactus garden[171] *which was one of the most famous in the world".*

After a while, Juanito invited Roger Martin to a hunting party near Madrid. Here, too, luxury was at its most. El Caudillo, Francisco Franco, was one of the guests. Roger Martin gives a lot of details in his book, but I don't want to deviate too long from our theme... If you find this interesting, you should revert to Roger Martin's book.
He continues: *"In spite of all the apparent complicity and a few other meetings, I have never seriously discussed Iberit problems with Juanito. Talks 'between the young ones' lasted. It turned out that there was no solution but to sell the Valladolid factory to Uralita. I paid a visit to the President of Banco de Santander. He was very rude to me and inveighed against the defeatism of Pont-à-Mousson. The merger with Saint-Gobain permitted to save face. Pont-à-Mousson had entered among great Spanish industrials. Banco de Santander resigned to the inevitable and we had 'to sue our pants off by Uralita.' "We made a mistake, we had to pay for it".*
This is how Uralita, already well established in asbestos-cement with its factories in Cerdanyola del Valle, Sevilla, and Getafe, committed itself to become the exclusive manufacturer in Spain. The company built the factory of Alcazar de San Juan and purchased Rocalla from Castelldefels, and finally Fibrotubo from Alicante and Valdemoro. The only Ibertubo from Toledo seems to have resisted despite a kind of drowsiness that may be related to the same situation that Iberit faced years ago.
When the asbestos crisis emerged, Uralita made big efforts to survive. It benefited from its long-ago harmony with Eternit, and more recently with the Australian James Hardie. But competition with new materials for the building industry appeared too harsh. Factories were closed one after the other, including the last one in Valladolid. Machinery and business of the latter was sold to Etex Group, which transferred part of it to the plant in Portillo. The factory born with Pont-à-Mousson around 1960 became in its turn an industrial wasteland, till the day when the abandoned site be wanted by some investors.
One may have an interesting glimpse on the recently modern Euronit plant settled by Etex in Portillo, not far from the old and abandoned plant of Valladolid.

171 The cactus garden still exists. If you visit Majorca Island, don't miss it. It is worth going.

An article in *El Pais* dated October 6th 1993,
Signed by Miguel Angel Noceda explained: The March family withdrew from the group starting in 1997, but essentially kept an eye on the company through the bank, of which they were one of the major shareholders, if not the first. In 2006, the emblematic name of Uralita disappeared from the world of large industrial building companies.

Sri Lanka

One morning in 1993, when the phone rang in my office and the operator told me the call was from a Swiss company, I only knew by name, I answered with some excitement.
The engineer at the other end of the line explained that his company had sold a Hatschek sheet machine to Sri-Lanka and that they had to supply the corresponding felts. In my mind, Sri-Lanka meant Ceylon and *"Féérie Cinghalaise"*[172] *a* book that had had some success, which I devoured in my teenage years and recently read again. That memories still makes me dream. I was proud that our company's fame was sufficient to draw the attention of a well-known Helvetian company in the world of asbestos-cement.
Now there was no longer any time to dream, but to win a new customer and perhaps penetrate a new market, even if the small size of the country could not grant for any major developments. Nonetheless, we signed the contract. A few weeks later we forwarded the felts to Colombo, according to buyer's instruction.
In February 1995, I took advantage of a commercial journey to Thailand, to make a stop in the country where elephants were still at work at the time. I obtained an appointment and soon landed in the former British colony, independent since 1972. The manager of the factory offered me a friendly welcome, the first felt was on the machine, had already worked a few hours for trials and the plant was to be officially inaugurated the next morning. I was invited to participate, which, of course, I accepted willingly.
At around ten o'clock the next day, I was back to the plant and discovered that offices I had passed the day before had been converted into banqueting rooms, and desks covered with tablecloth, flowers and all kind of good food, watched by women whom European clothes from the day before had been replaced by smart "saris". Mood seemed pleasant. The manager of the factory informed me that the president of the company, the Anglican bishop, who would bless the machine, and local VIPs, would be arriving within minutes. I inquired about taking snapshots, the answer was negative. Consequently, I could not bring souvenirs to the family.

[172] *"Féérie Cinghalaise"*: (could be translated as" Wonder Ceylon") by Francis de Croisset 1928 about Sri-Lanka when" British Empire" had a true meaning.

Authorities arrived and after usual congratulations they climbed on the stage right in front of the control desk, next to the making roll. The president made a large speech in his native language. I concluded that it was a very good one as people applauded heartily.
Then it was the turn of the bishop, who read nice words in a large book while the president, who had become altar boy in between, picked up a large holy water stoup from which the sprinkler handle protruded. The bishop blessed the control panel and the employees standing nearby. Then, he and the president as an altar boy began walking the whole machine, blessing each part as the crowd followed them. They walked with large steps, blessing anything that seemed them deserve such an honour. A machine of the type and its accessories are quite long, maybe 80 or 100 meters. When they got to the end of the Hatschek machine, since the holy water stoup was not empty, they opened a door that gave access to offices and carried-on blessing tables, chairs and even the telex machine. But since there was still holy water left, the bishop opened an anonymous door and blessed the toilets with a large and final cross. That was the end of the blessing of the Swiss machine and everyone hurried to the fruit and cake that waited quietly for us.
I have to believe the blessing was effective as our felts performed well and we received more orders in the years that followed.

More recently, the Sri-Lankan revue, Mirror Business, reported on October 21st, 2016 about the activities of the Russian Ambassador in his country.

Zahara Zuhair, quote:
"Ambassador of the Russian Federation to Sri Lanka, Alexander Karchava, said the postponement of the proposed ban on importation of chrysotile asbestos till 2024 does not resolve the problem, and from their side they are ready to bring in a group of experts to Sri Lanka to educate the people on the safe use of chrysotile asbestos.....
...The ambassador urged the members of the Sri Lanka-Russia Business Council to express their position on this matter and use their influence to address this matter, which is actually not only in the interest of Russia, but also to protect the rights and profits of their countrymen.
The ambassador said he personally discussed the issue with President Maithripala

Sirisena and Prime Minister Ranil Wickremesinghe. He said Wickremesinghe assigned Law and Order Minister Sagala Ratnayake to address this issue."

On another note, I told you "that in 1995 I was not allowed to take pictures when the new machine was blessed". Fortunately, times have changed and I can now show you a contemporary Hatschek machine that Wehrhahn set up at El Toro's premises in 2017 for the production of autoclaved flat sheets. Another one has been ordered and will be delivered pretty soon.

08-043 Wehrhahn Hatschek Machine at El Toro's premises

Switzerland

As already mentioned , Switzerland has taken first place in the fibre-cement industry. In 1903, Industrial Alois Steinmann, founded *Schweizeriche Eternit AG* and the factory in Niederurnen, Glaris, followed soon by a second one in Payerne, Broye-Vully, Vaud.

Throughout the 20th century, the company grew, changed its name, developed abroad with the establishment of branches and in association with the other Eternit family members, participated in large international fairs, sometimes suffered very serious setbacks that threatened its future, but after all, Eternit Switzerland always survived.

A recent proof can be found in 2017, with the arrival of:

"Swisspearl® FLOOR, presented as a prominent product, wished and expected by so many architects.

Conversations and wishes have been the subject of discussions for decades. Since last March we can finally provide an answer: Eternit fibre-cement, which was used until

now in fields such as façades, roofing, as well as in 'Garden & Design', is now available in previously unknown qualities.
A conventional fibre-cement surface presents a filigree texture. The peaceful feeling of a unique and natural new fibre-cement, offers a light range of colours. Authenticity in esthetics and touch are always warranted thanks to the first-class coating of the sheets. The colour palette in itself is a wonder."

08-044 Swisspearl floor. Company's Website

This was more or less the laudatory text that appeared on the Swiss Eternit website when I was preparing the French version of this book. It now seems less eulogistic. Contemporary creations and fulfilments are in no case limited to the above. Let us have a look to some of them:

-specially designed seats for motor-ways rest area,

-**Avera,** is a new fibre-cement sheet façade with: *"living, authentic and natural texture. That opens doors to surprising possibilities of highly expressive façades"* explained the inventor.

-Soundproof walls for motorway side.

We only talk about fibre-cement innovations, but the Company is also very involved in diversification.

08-045 Swisspearl soundproof wall for motorways. Company's Website

Eternit Switzerland is the living and visible driving force of the fibre-cement industry in the Helvetian Confederation, some others, less known from the public, intensely participated to the wealth of the country. This has been the case with asbestos-fibre trade for decades. Even though asbestos bags were not transported on the side of Lake Leman, banks and trading organisations were happily located around. One can imagine that such a business has become less active since the general European asbestos ban.

Another area in which Switzerland was very efficient in the fibre-cement world were machines and accessories that are indispensable for the production of sheets. I told you about my contact with a machine builder who gave me the opportunity to see the "show-like" blessing of a machine in Sri-Lanka, but machines of the same builder made me travel through many Latin American countries where I could, very often, meet the same high quality machinery that benefited of the last technical improvements of the time. It is useless to say that my national chauvinism would have been happier if such tools had had Asterix' country for origin.

08-046 Moderns Swiss Hatschek machine. Website of now disappeared machine builder

Before we leave Switzerland, let us praise one of my sources, I mean: "*Bulletin Technique de la Suisse Romande* "which has already been quoted, and especially an article by Mister H. FREY, Engineer Manager of Eternit in Niederurnen, which provides valuable information about the development of the company he manages.

Syria/Lebanon

After long preparation with help of a Syrian gentleman living in Paris, a prospecting journey to Lebanon and Syria was conceived in January 1994. We had received inquiries for large quantities of pipe felts from a Syrian company, but we did not want to take the risk of such an order without knowing what exactly the technical conditions were. The second or maybe the third Lebanon war was just over. However, my Parisian mentor insisted on keeping my safety in the best possible circumstances. This made me live a VIP treatment. At Beirut airport, a microphone call invited me to report to the nearest crew member as soon as the plane stopped on the tarmac. I was the first passenger to get off the plane. Right under the wing two bodyguards and a limo expected me. When I was in the car, one of the bodyguards asked for my passport and luggage ticket, then I was driven to the VIP lounge, where I was offered some champagne. Half an hour later, my angels were back with my passport and my suitcase. We got in the car and drove to my hotel. There they watched whether my check-in was OK before they left, renewed their advice not to leave the hotel and informed me that they would be back around 10 a.m. the next day. I was condemned to stay in the hotel. The complete opposite of my usual practice, as I like to start my visit to a new country with a walk, buy a drink and a local newspaper if I am able to understand the language, and have dinner in a local restaurant.

The next morning at 10 a.m., one of the guards was behind the wheel of a shiny Mercedes 600. I soon found out that it was armoured. He took me to a beautiful villa where I was welcomed by an elegantly dressed middle-aged man. I immediately discovered that he had no idea as to what my visit to Lebanon was for. I did my best to inform him, but it was clear that he was not interested. We held our conversation until lunch time. We drove to the restaurant in a white Mercedes 600 - the previous one was black- driven by a new chauffeur who was waiting for us in the garden.

The meal was certainly a rich lunch, but painful for me and probably for my host as well. We went and wondered what our meeting was for. The only positive point was that I had an appointment for the next morning with the manager of the Eternit local factory. My interlocutor had promised to organise the meeting and those foreseen in Syria for the coming days, the main reason for my presence.

The afternoon was free, and I intended to forget about the guards' advice "not to go out". Unfortunately at that time Beirut had no point in common with that of my dreams. I had never seen such a mess of electric cables, *(later on I saw the same type of mess in many places)* assorted of very frequent breaks of light, even during WWII. But what most impressed me, was the extraordinary number of ruined buildings, and the number of holes due to shells in the broken remaining walls.

The next day, the visit to the local fibre-cement factory resulted perfectly useless. The manager was nice, but he did not need anything and said he was perfectly satisfied with his usual felt suppliers. Moreover, activity was so low that renewable parts were granted a long lifetime.

In the afternoon, the planned meetings in Syria and journey plan were confirmed. On the third day around 8 a.m. the bodyguards came for me with a big 4 x 4 equally armoured and we headed to Damas. Everything went well, the technical manager of the Damas pipe plant, supposed to be interested in our products and references, provided me with the necessary technical data to make an offer possible. After a typical lunch in a small and pleasant place together with my chauffeur we took the highway to Aleppo. It's been quite a long journey and my first surprise was when the chauffeur pulled a KA 47 out from under the dashboard. He was proud of his tool and ensured that with such an angel we did not risk anything...A second surprise occurred right after when I saw a lorry coming towards us, on our own lane, on the motorway, followed by a second one and others... I expected a strong reaction from my driver, amazement, swearwords or anything else...but none, more trucks and a few private cars were passing in the opposite direction, while the lanes on the other side of the central reservation were pretty empty except for a few vehicles heading the same direction as we did...By luck, traffic was slow and it was clear that local motorists used the four lanes in any direction without caring about common rules...and it was considered as absolutely natural. I never went back to Syria, so I could not check if the rule changed...Donkey carts were also allowed on the motor way, it seemed perfectly legal.

When I suggested the driver to have a stop for a drink, foreseeing something on the other side that could be a bar, he did not hesitate and the big wheels of the 4x4 were very welcome to cross the big ditch between the two ways.

It must be noted that he proceeded equally when we left the bar...Like so many other drivers, he could have driven and stayed in the lane. We left the motorway at the next intersection. I wondered why: simply asking a shepherd who was looking after his skinny sheeps if we were on the right way to Aleppo. The same was repeated twice before we reached our destination, now a martyrized town. My conclusion was: "AK 47 has become a daily tool, but road maps are not".

We finally reached Aleppo and our five star hotel in the middle of the afternoon.

A few days before I left Angoulême, Charente, France, a young colleague had heard of my upcoming journey to Syria. He was married to a Syrian woman and asked me if I would accept giving a few dollars to his step mother living in Aleppo. I accepted the task with pleasure. So I had an envelope in my pocket that contained two or three hundred USD and the phone number to call. As soon as I was in my room I called. A few minutes later, I was informed by the front desk that a lady was asking for me. A young man was there with the lady, they both thanked me and, probably as a reward for my "good deed", I was told that the family was going to celebrate the 15th birthday of a young girl that evening: "*Would you join us*?" Of course I was tempted to accept, but it was intended that I would have dinner with the driver. *"The driver will be welcome…".* I asked him and he mumbled that he had planned to meet someone after dinner…Consequently, I felt free and we agreed about the time to pick me up.

Around 8 p.m., a new car brought me to a pleasant and cosy three-story villa, in which a large number of people were already gathering in reception rooms. The introduction seemed a bit difficult due to language problems, but when I mixed English and German I understood that I was in a rich family of surgeons who shared their time between Aleppo, Cairo and some major European towns…

After a while the "15 year old girl" was announced. She was dressed in white and wore a light white veil -which did not hide her face- for hazy memories I cannot tell you if she was pretty. I only remember that she was somewhat plumply.

Then we sat down for dinner and there were many different courses on the table, but I had to accept the local rules: eat without a plate, carefully place a small piece of one's choice on top of what could be bread if cooked, and then eat without scattering the pieces…

At the end of the meal, two young men took me back to the hotel. They drove slowly through the old famous town, offering me a short sightseeing. Years later, I still wonder why my French colleague deprived himself of his dollars to the benefit of a lady whose family income obviously was much better than his own. I also wonder how, after one single night of staying in Syria, I could deserve the privilege to participate in a "15th birthday celebration" of a young Alepine heiress?

The following morning, the visit to the pipe plant offered me the same kind of meeting that I had known in Cairo a few years earlier. I mean, if any European industrialist wanted to penetrate the market, he had to foresee one or more discreet, well led envelopes to be handed to the right person. I heard the rule resulted from the fact that company executives did not receive the wages they deserved.

In conclusion, I should say: even if the journey gave me many personal memories, it did not bring any profit to the company who paid me. As a matter of fact, slipping envelopes was not part of our working procedures.

T...for

Taiwan

In 1987, a fibre-cement department was created under the name of Douglas in Taipei, to produce ceilings and dry walls. Asbest-free technologies were used immediately.
Douglas claims to be present in the Middle East, Southeast, United States, Latin America, North Africa and Europe markets.
The two images below, taken from the Douglas advertisement, clearly show what is required to make fibre-cement available for all types of buildings.

08-047 /048 Two pictures out of Douglas Website

Thailand

Thailand is a very important fibre-cement industry. This will certainly not surprise the traveller who has come through Bangkok, where he has seen the huge development of the last 50 years.

Thai Olympic Fibre-Cement

In 1974 Thaï Olympic Fibre-Cement Co Ltd was founded, which today employs 200 people and is to have its main markets in the following regions: North and South America as well as in Eastern Europe. The company produces corrugated sheets, flat sheets, floors of different types, some of them without asbestos as stated on its website. Quote:

"These products are from our own experience and expertise in this field with world-class autoclave technology, via non-asbestos fibre-cement product".

Siam Fibre Cement

Siam Fibre Cement, member of the large group Siam Cement Public Co. Ltd. (or "SCC"), was already a big producer of fibre-cement when I sold them felts in the late the 20th century. In 2010, they announced a huge investment permitting to produce 22 million square meters more each year, a total of 135 million a year. Why? Just to meet the market demand...

Mahaphant Fibre Cement

Mahaphant Fibre Cement Public Co Ltd has also done me the honor of being one of their felt suppliers. Since then, they have been able to switch to modern asbestos-less fibre-cement, for instance panels and sidings, although producing on double width, even on triple width Hatschek machines.

More recently, their brand SHERA has developed so-called "ply"[173] products, which range from ceilings and walls to furniture for kitchens, bath-rooms and laboratories, etc. Doesn't it remind us of the famous Eternit Glasal?

[173] "Ply", very often associated with "plywood" lets think that we are talking about fibre-cement imitation of plywood. If a reader can confirm or explain this, they are welcome.

Mahaphant owned several plants in Thailand when I knew them, but they now appear as Mahaphant Fibre-Cement (South Asia) Pvt. Ltd. headquartered in Maharashtra 400053, India, and according to the number of sites on their internet website, it is self-explaining in terms of their growth, which undoubtedly corresponds to the general demand in building products in Asia.

08-049 Sight of Lopburi plant, one of Mahaphant five plants in Thailand

08-050 Fibrecement Thai house. Mahaphant Web site

Private anecdote:
The first time I visited a large Thai group in Bangkok, the appointment was set at 7 a.m., Understand: 2 hours between the hotel and the customer, intended for by the doorman, plus one hour to get ready and have breakfast, which meant getting up around 4 a.m., without considering the jet lag and time difference.
The buyer, a charming young lady, welcomed me, and we drove together in the unavoidable VW company combi in hands of an experienced driver to the plant about 30 km away. After the technical visit, I found out on our way back that my representative was learning French and had heard of the famous *"Le Petit-Prince"*[174] from Saint-Exupery. A few months later, when I came back, I brought her the famous book and the corresponding CD, recorded by the also famous Gerard Philippe. Of course, she was very happy with the small gift.
The following appointments were organised according to more convenient times for European landing in South-East Asia. I never knew whether the gift would be part of our good business relationship, since, invited to retire, I no longer traveled to the country of so many buddhas.

Turkey

Founded in 1956 "Atermit" was the first enterprise to produce asbestos-cement in the country. It was a quick success, especially in the field of corrugated sheets, which has definitely become the favourite roofing for countless buildings.
When asbestos was banned *(December2010)*, Atermit did not wait long to turn to new technologies. Their website informs that their workshops have produced more than 150 million square meters since the company was founded, but does not give the percentage between asbestos-cement and NT.
In 2001, Oner and Nafiz Hekim founded "*Hekim Yapi Endustrisi Sanayi Ve Ticaret AS*" affiliated to Hekim holding in view to produce basic cement sidings. They chose fibre-cement as the most widely used in the world and began manufacturing sheets using the new technology.

[174] *"Le Petit Prince"* The Little Prince by Antoine de Saint Exupery, the most famous work of the author, who in the 1930s was also one of the founders of airmail between Europe and Latin America. He died when his fighter aircraft was shot down in the Mediterranean sea at the end of WWII.

In 2007, without waiting for the asbestos ban in Turkey, the founders acquired the "SIM s.r.l." an Italian licence and started a new production line in 2008. Their own investment funds enabled the production of around 80,000 sqms per year in their premises in Adapazari Hendek[175].

[175] Information that can be found on websites varies essentially depending on the date of the visit.

U for

United-Kingdom

It seems that the British were interested in asbestos long before the arrival of asbestos-cement in their country. Cape Asbestos was founded in 1893 with the purpose of producing asbestos fibres in South Africa. Initially it was a large consumer of the fibres in textiles and bituminous cardboards. Much later, they focused on asbestos-cement. *(We could see that James Hardie discovered asbestos-cement in France and not in his own country, while travelling around Europe at the beginning of the 20th century.)*

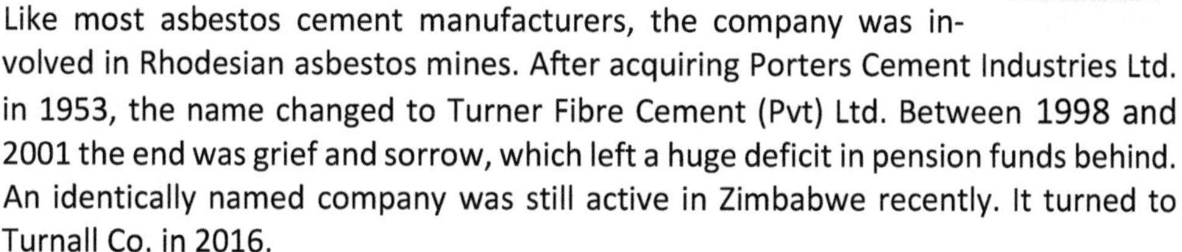

Turner & Newall
Shortly before the First World War, the Turner Brother Asbestos Company, which had long specialised in asbestos weaving, opened an asbestos-cement factory in Trafford Park, Manchester. There, they manufactured asbestos-cement "Trafford Tiles" mostly used in roofing of agricultural and industrial buildings.
In 1920, after a merger with Newalls Insulation Co., they became Turner & Newall. Soon after, they were listed in the London Stock Exchange. During President Samuel Turner's years, the company was involved in an Industrial Administration School and a Dental School in Manchester. It did not stop it from developing and acquiring Bells United Asbestos Co., as well as several other companies active in the asbestos-based insulation field. When they closed the factory in 1959, the "Armley Asbestos Disaster" occurred in relation to a large area comprising more or less a thousand houses built on asbestos-cement sludge.
Like most asbestos cement manufacturers, the company was involved in Rhodesian asbestos mines. After acquiring Porters Cement Industries Ltd. in 1953, the name changed to Turner Fibre Cement (Pvt) Ltd. Between 1998 and 2001 the end was grief and sorrow, which left a huge deficit in pension funds behind. An identically named company was still active in Zimbabwe recently. It turned to Turnall Co. in 2016.

Cape Asbestos

In 1953, Cape Asbestos associated with the American John's Manville, to produce Marinite, also referred to, at its factory in Uxbridge, Middlesex.
In 1960, the company acquired Universal Asbestos Manufacturing Co. Ltd., and entered the world of asbestos-cement. Around 1973, it bought John's Manville shares and around 1980, it merged with James Hardie, abandoned asbestos and adopted its new partner's technology for producing autoclaved cellulose-cement.
The result was the birth of two entities:
- "Cape Board and panels" that produced light panels on a Hatschek machine in its Uxbridge factory:
  - -Masterboard for internal use,
  - -Masterclad for external use.

According to some French specialists, the company had perfectly mastered the technologies.
- "Cape Unicem", whose factory produced corrugated sheets in Bowburn, near Durham, West Sunderland, also abandoned asbestos and came up with compressed autoclaved cellulose-cement *(with a Simpelkampf press)* using James Hardie technologies. The specialist to whom we owe the specifications revealed, around 1985, to have seen poor quality sheets and moldings in Bowburn, that were very different from those produced by James Hardie in Australia. The Bowburn factory closed in 1989 and was demolished soon after.

Cembrit
The Danish company is present in the UK, but does not appear to have production in the country.
Let us digress for a while and talk about the quality control system that the city of Birmingham developed in the late 1980s. Probably due to some fraudulent quality problems resulting from asbestos-free artificial slates recently placed on the market, the city council envisioned an evaluation test for such slates. It became a reference, first for the council, but soon for other big British cities as well.
The test was both simple and fearful for industrials of the field. Called "Heat-rain test" and consisted of a small 4 to 6 m^2 roof model exposed to heat cycling to 60/70°C (140/160°F) using infrared radiation, followed by strong spraying of water at ambient temperature.

Slates were set English style, that is, with two nails and a wind clamp. Few suppliers passed the test. The French Everite obtained a very good result against its British competitors. Unfortunately, for the French company, success was not followed by its sales department and consequently did not produce the possibly expected results.
According to our source, the control organism of Birmingham is still active, and it is fair to believe that the slates from the Eternit factory in Saint-Gregoire, Brittany, France will pass the test as they are steadily supplying the British market.
Almost ready to close the UK chapter, I discovered the existence of "Fibre Cement Guttering" stockiest of Marley Eternit UK, which offers its products *"the modern-day replacement to asbestos cement guttering"*.
As I questioned the company about the manufacturing process of products like those hereunder, it was confirmed that they are actually manufactured with flat sheets coming from Hatschek machines. Sheets are then molded in order to obtain the desired products. We are right there in the fibre-cement process described in Chapter III," Manufacturing Process and Products". The manufacturer stated that most products are made in a very short time. Some of them, such as "Swann Neck", could take up to two months to be molded in the right shape and dimension. Such a delay seems to justify the numbers appearing on the price list. *(£ 145,13 for a -150mm, 250mm projection -Swann Neck August 2019).*

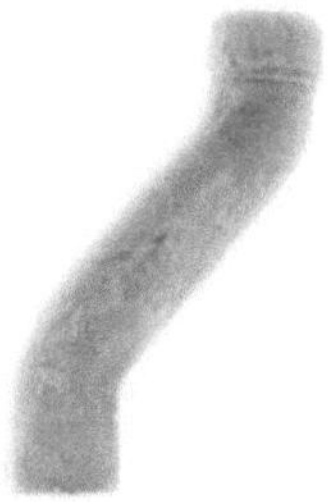

Swann deck

Downspout

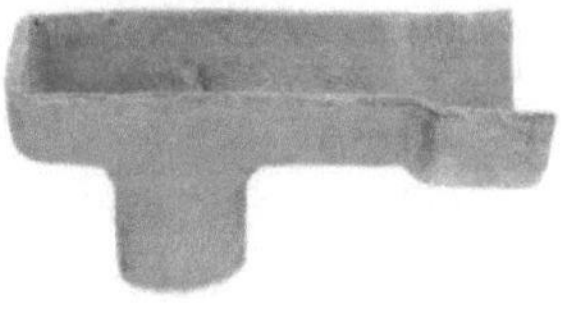

Stab connection

USA

In the case of USA it is for sure very difficult to deal with as it includes everything, from research of new building products at the end of the 19th, to the US market desire from every fibre-cement group in the world, as well as health scandals with all the grief that comes with it. At that point, I am tempted to write:
"The field is so wide and laborious that it would justify a whole book in itself. If any reader feels like undertaking the task, I gladly let him do..."
Finally, moral duty makes me try the operation, pointing out the main points, or at least those that I consider as such.
First, one famous name stands out across the North American continent, and even beyond that I mean Johns-Manville. Here is one of the oldest ads, published in 'The Tech Boston, Mass.' dated April 14th 1920.
One has to admit that in 1920 Johns-Manville was no longer a "start-up". The number of cities it was commercially located, as you can read in the present ad, should be sufficient to convince us.

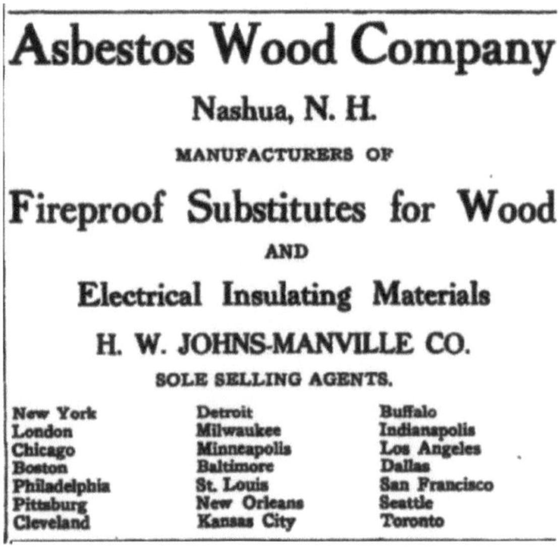
Asbestos Wood Company
Nashua, N. H.
MANUFACTURERS OF
Fireproof Substitutes for Wood
AND
Electrical Insulating Materials
H. W. JOHNS-MANVILLE CO.
SOLE SELLING AGENTS.
New York
London
Chicago
Boston
Philadelphia
Pittsburg
Cleveland
Detroit
Milwaukee
Minneapolis
Baltimore
St. Louis
New Orleans
Kansas City
Buffalo
Indianapolis
Los Angeles
Dallas
San Francisco
Seattle
Toronto

08-053. H.W. Johns-Manville CO. The Tech Boston, Marsh 14 1920.

Founded in 1858, as H.X. Johns Manville Manufacturing Company, already used asbestos fibres in the manufacturing of fire proof products. This justified its interest in ASBESTOS, Canada, town already largely quoted earlier. In 1901, it had merged with another company of a similar name "The Manville Company" from Milwaukee, operating in the same field of activity. This became the: "H.W. John's Manville Company". Soon they included the newly born asbestos-cement to their range of products. During the two world wars, which in the meantime became the "John Manville Corporation", the company took an active part in efforts that were required for military needs.

In 1945, the US government invited Manville to invest in research for fireproof products required by the Navy. The result was new products that mixed asbestos and silica. *(It may be interesting to note that several decades later, old civil ships and warships were tugged to Asian countries for disassembling regardless of the workers, which led to controversy and scandals of all kinds).* During a large part of the 20th century, the company was a leader in asbestos-cement products such as ventilation and water pipes, as well as shingle and roofing sheets.

As early as 1958, perhaps feeling the change of wind, the company pursued its diversification and became involved in glass-fibre.

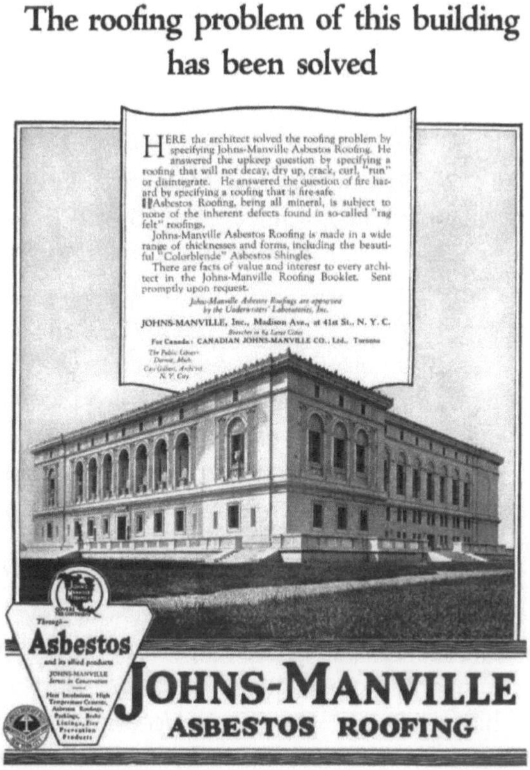

Beginning in 1982, thousands of people began developing asbestos-related illnesses that led to lawsuits against the company and eventually filed for bankruptcy. It overcame bankruptcy in 1988 and founded then the "Manville Personal Injury Settlement Trust" granted with huge funds, *"to compensate people suffering personal injuries from exposure to asbestos or asbestos-containing products manufactured or sold by Manville"*. Casualties could be financially compensated and shareholders could feel better.

In 1997 an important event occurred, a new name change: the dash between John and Manville was erased. The company had its previous name again: "John Manville".

In 1998, a new plant was built in Macon, Georgia, that had been almost invisible for twenty years. At the same time, in Jersey, a town where the company had long owned a very large plant, took the name Manville.

CertainTeed

In 1904, another future building industry giant was born in Saint Louis, Illinois. It was called "The General Roofing Manufacturing Company". The founder, Georges M. Brown, started the business with $25,000.

Thirteen years later, on the opportunity of a reorganisation, the company chose a new name based on its own motto: "*Quality made **certain**, satisfaction guaran**teed***". The next year, CertainTeed was listed in the New-York Stock Exchange.

Among the numerous activities related to construction and infrastructure, CertainTeed developed a large production of autoclaved fibre-cement pipes over the years.

Around 1960, the company acquired the "Asbestos-Cement Pipe Division" of the Ambler-Pennsylvania-and-Mattison Company[176], second largest American manufacturer.

In 1967, the French Saint-Gobain owned a small part of the company's capital.

In the early 1970s, CertainTeed's wish to develop a glass-fibre activity and that of Saint-Gobain[177] to increase its position in the US building market, encouraged the two companies to become closer. Roger Martin, then President of Saint-Gobain told us in his memories: "*The offer, which was made on Wall Street on May 1st, 1973, was, I believe, one of the first ever accepted by a French company. 4 million shares were offered, we took 2,500,000 and with Mexalit's contribution our stake was increased to 34%*[178]". A few month later, the stake was increased to 40%, soon passed 50% and finally reached 100% thanks to the purchase of the "British Turner and Newall", who wished to leave the field.

A retired executive of Everite France R&D, who took a business trip at CertainTeed in 1985, told me recently: "*I visited the autoclaved pipe plant of Hillsboro, Texas, together with the manager of the department. I was impressed by the cleanliness of*

---

[176] The Ambler-Pennsylvania website now appears to be inaccessible.

[177] In the meantime, Saint-Gobain had merged with Pont-à-Mousson thus took over part of the fibre-cement activities in France and in some other countries.

[178] Patron de droit divin, already quoted by Roger Martin, page 351.

the plant, which still used the asbestos willow[179] *drying process (which had already been abandoned in Europe). I don't remember anything specific about the technologies, but what I was most impressed with was the manager who went nowhere without his Texan hat and six-shooters. Let us remember that we were very close to Dallas..."*

CertainTeed could not escape the asbestos crisis and suffered severe attacks from casualties of the fibre. In 1990 it was decided to completely remove asbestos. Thanks to its other activities CertainTeed remained present in the all-important building industry in the United States.

Six years later, in 1996, CertainTeed probably driven by its vinyl siding success and huge local demand, celebrated with the help of European technology and the assistance of Saint-Gobain. The general management in Valley Forge, Pennsylvania surprised every competitor by announcing that this time around, it would return to the fibre-cement market without asbestos.

In 1998, the newly created plant in White City, Oregon began using the latest European technology. The plant produced wood-looking-like NT planks and siding with unique characteristics at the time that offer smart textures with matching shades and seduce both residential and commercial buildings, as can be verified on the website.

08-055/056 Two examples of CertainTeed are offered on its website

[179] Willow: Hammer mill working with dry process. *08-055-056.*

To strengthen its market share, CertainTeed acquired the factory of Roaring River, North Carolina. Only two years after the building by ABT Co., entry into the market had failed.

In 2006, CertainTeed attempted a third plant in Terre-Haute, Indiana, but the results were not hopeful, likely due to the sadly famous subprime crisis that hit the entire US economy, and the building industry.

At the same time, CertainTeed received a purchase offer from the Mexican Elementia, already referred to in previous pages. Finally, in 2014, that is only 14 years after resuming its fibre-cement activities, the company sold them to Elementia. The latter united them with that of Allura, another American group that was already in the hands of the Mexican giant.

When the great asbestos crisis described in chapter VI struck, the parent companies of the US branches knew full well that they would be risking a lot. One has to keep in mind that in the US, unlike France and some other countries, compensation is not required to prove the damage. The fact that a worker claims to have come into contact with the occupational risk is enough to initiate legal proceedings. Before taking the risk, most managers decided to negotiate a transactional indemnity through their lawyers.

This explains the large number of litigation still going on today. In 2016, they would still number in the thousands.

As early as 2002, Jean Louis Beffa, President of Saint-Gobain, *(for the record, Saint Gobain is the parent company of CertainTeed)* decided to provide € 100 million against the

risk for the current year only. The case was confirmed by "Le Moniteur" [180]on July 26th, 2002 quoting the words of the President: *"In a way, Saint-Gobain will become its own insurer"*.
Some ill-speaking people then predicted a probable descent into hell for the three hundred year old French group. The following years and its 350th birthday in 2015 clearly proved that they were wrong.

Allura

Founded during the Second World War, Allura mainly produced outdoor fibre-cement such as shingle, panel, rough thick slate, frame and siding. Allura was handed over to Elementia in 2014 and operated three plants that were retrieved by CertainTeed:
-White City, Oregon,
-Wilkesboro, North Carolina
-Haute Terre, Indiana.

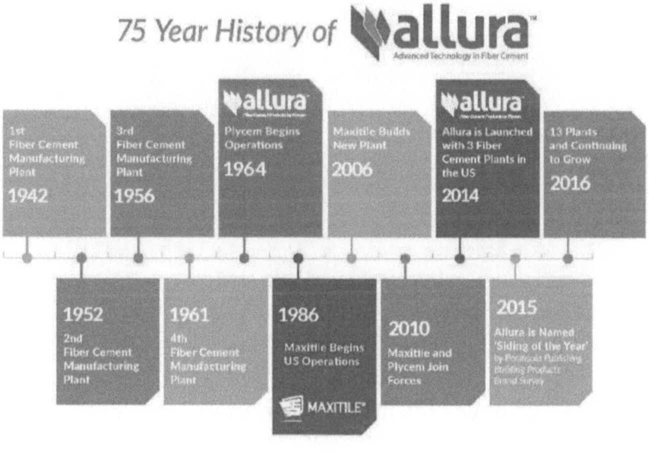

[180] *Le Moniteur* is a first-class French weekly magazine dedicated to building and quality of life in France.

08- 059 Allura : Contemporary Home- Richmond

The name of Allura,[181] is also associated with Plycem, which originally comes from Costa Rica and was founded by Stephan Schmidheiny after he left Eternit Switzerland...

[181] At the time I was writing this English version of the book, Allura's website answered strangely: *"The owner of this website has banned the country (FR) of your IP address from accessing this website"*.

American Fibrecement Corporation

Denmark gave us the opportunity to talk about Cembrit and Russia. Remember, the pipe machines that were ordered by the Tzar and never delivered for reasons of war and revolution...
In France, we use to say marmalade attracts flies, but we can also say that the US market attracts everyone, including fibre-cement manufacturers.
After Saint-Gobain, this was the case with Cembrit and James Hardie.

Cembrit made a deal with US investors and in the late 20th century founded the American Fiber Cement Corporation, well known as AFCC, using their technology.
AFCC bases part of its marketing on the European technology acquired with Cembrit and insists on:

"Fibre cement cladding panels are ideal for ventilated, light weight façades, including wall systems, soffits, façades, window elements and balcony boards - just to name a few. When you select fibre-cement cladding panels for your building, you benefit from numerous advantages and flexibility. Fibre cement is sustainable due to its durable nature. High density, compressed fibre cement has excellent mechanical properties with superb freeze-thaw characteristics and resistance to fungus, mould and bacteria. These virtually maintenance free materials combined with the ventilated, insulated rain screen construction contribute to increased heating and cooling efficiency. The panels come in a range of colours to help you realise your architectural vision. They are easy to handle and meet the need for ease of installation with our technical support and supply of all required accessories".

Every manufacturer in the world seems to repeat the same arguments. On this occasion, Cembrit reminds us of the components of modern fibre cement. We can read as follows in Cembrit Corporate brochure:

Our recipe for high-performance fibre-reinforced cement products:

-CEMENT (65–80%) Mostly a mixture of burnt limestone and clay that sets and hardens when exposed to water.
-PVA (2%) Engineered polyvinyl alcohol (PVA) fibres for reinforcement and to increase product strength, durability and flexibility.
-FILLER (10–25%) Mainly passive (i.e. non-binding) materials (limestone, micro silica and recycled fibre cement, etc. to make the mixture easier to process or to add specific properties.
-CELLULOSE (3–5%)Wood fibres help strengthen the product as well as keeping it flexible, but are mainly used as a production process aid.
-WATER The rest of the mix is water, added to get the process started.

When I first visited the AFCC website, the products were granted a *"50 year warranty, regardless of climate"*. During my new visit in August 2019, I found a mention of only ten years, which sounds a bit more *like the time...*

James Hardie

You have heard of James Hardie on the pages dedicated to Australia, the company's original country, and some Southeast Asian countries as well, but his part in the US market, surely deserves a few more lines.
In the years 1987-1988 James Hardie began to import *"Hardie-Shake"* from Australia. *"Hardie-Shake"* is autoclaved fibre-cement with a wood aspect, inspired by the conventional wood-shake that is so widely used in Uncle Sam's country. The first time it was a small size issue: 6in, 8in, and 12in wide and 24in long. At the same time, the conventional US wood-shake began to lose market share due to the fire risk and the resulting increase in insurance costs. James Hardie first started a small packaging workshop located in Fontana, California. Soon fire resistance and the arrival of new shades brought success, even if no one knew the ability of the product to resist the Western US climate in the long term. Such a success contributed to a growing distrust of wood-shake, pushed James Hardie to turn his Fontana workshop into a real factory, capable to produce fibre-cement "Hardie-Planks" up to 7.25 inches wide and 12 feet long.

The growth in sales has attracted other world producers to the US plank market. But over time, fungal infestiations appeard, particularly aggressive in humid climates, which brought the first trouble. It was then clear that autoclaved cellulose-cement sheets, both flat and corrugated, could not exceed the height or the ten year guarantee.

Some foreign fibre-cement manufacturers proposed non-autoclaved products including polymer fibres with a certified warranty of fifteen to twenty years. James Hardie refused to play that game. It should also be noted that roof products are always more at risk than vertically used façade products due to their direct weather influences. The latter, which are painted, benefit from additional protection.

The strong market demand pushed local companies with no experience in the fibre-cement industry to start an alternative production of European origin. The success was not always there and the demands multiplied. Disappointments from unproved newcomers led them to sell their factories, which were soon purchased by experienced fibre-cement makers, including James Hardie. In this way, James Hardie, who was already a leader in the US market, was spearheading height plants in 2017.

One can list the following factories without knowing the exact date of each opening:

-Fontana, California, the first built, primarily for R&D.
-Reno, Nevada, with 123 workers. Was the first to recycle and reuse its own water.
-Plant City, Florida, with 156 workers
-Peru, Illinois, with 290 workers
-Cleburne, Texas, the largest southern plant, with 249 workers
-Waxahachie, Texas, counting 174 workers
-Pulaski, Virginia, the largest of them all, with a total of 350 workers
-Tacoma, Washington, with 174 workers

In California, Mission Viejo focuses on accounting, human resource, analysis and marketing.

Headquarters are located in Illinois, Chicago and Naperville with many central services such as financial analysis, strategic marketing, price studies, recruiting...

In May 2017, the company was delighted and proud to announce the creation of its ninth American plant in Prattville, Alabama representing a 220 million dollar investment with an estimated 205 jobs and better service to southern customers.

James Hardie reports on one of its marketing campaigns on his website:

"We innovated a localized approach to siding with 'The HardieZone System' of siding products in the mid-2000s. All James Hardie siding products are Engineered for Climate®, meaning they are formulated to resist damage from weather conditions in one of two distinct regions of the United States. The HardieZone System is based on eight individual climatic variables that primarily affect long-term performance of siding. Using these factors, the company arrived at 10 distinct climatic zones, with common variables in certain regions, allowing it to engineer the HZ5® product line for roughly the U.S. Midwest and Northeast, where snow, ice and temperature swings are the norm, and the HZ10® product line for roughly the U.S. South and Southwest and its prevalent heat, humidity, sun and rain"

Despite the technological progress known in the 20th century, colour remains a major problem for fibre-cement manufacturers:
In the case of "plank siding" the primary colour, the final colour and the wide range of available and required shades give those responsible a real headache. James Hardie seems to have overcome the difficulty, as stated on his website which can be read:
"THE COLORPLUS® TECHNOLOGY WARRANTY
ColorPlus® Technology finish comes with a 15-year limited warranty that covers paint and labor, protecting against peeling, cracking, and chipping".
Nevertheless, it happens that the company suffers attacks from unsatisfied buyers. But nobody is perfect in our world!
As in any industry, James Hardie lay low and waited until the market tended to decline for economic reasons. This was the case in 2008 when 60 people were laid off in one plant and 67 in another
Fortunately for enterprises, as well as for the many people involved, future prospects seem to be quite favourable, as Cision Institute PR Newswire told us in 2016:
"US demand to rise 6% annually through 2019 to 2.9 billion square feet, valued at $2.2 billion. Rebounding construction expenditures, specifically in new housing construction, will fortify demand. This growth in new housing construction will be by far the biggest driver of fibre cement product demand growth, as single-family housing starts advance rapidly and fibre cement siding -- the most popular application for fibre cement -- continues to gain market share nationally. Growth in fibre cement siding use in single-family housing will be supported by fibre cement's high levels of market penetration in the south and west, the US regions that will experience the strongest levels

of growth in population and housing starts going forward. ...Homeowners can engage in renovation activities like bathroom and kitchen remodelling, which will support demand for fibre cement backerboard[182]*. Even when properties are not changing hands, fibre-cement products will make inroads in re-siding and exterior trim applications".*

[182] *Backer board*, or CBU, is a mineral-based board that you can attach to your wall if the surface is not suitable. You can tile directly on it.

V for

Venezuela

In 1946 a well-known Venezuelan industrialist founded Canacit in Caracas with the help of the French Pont-à-Mousson.
In 1975, the Venezuelan government wanted factories to move away from the capital and offered financial compensation to companies that responded positively. In this way, Saint-Gobain, who had in the meantime merged with Pont-à-Mousson, closed the already old factory of Caracas and opened a new, naturally more modern one, about 80 kms from the capital, in a place called San Francisco de Yare.
Two engineers of Everitube spent some time in the new plant. My encounters with them give me the opportunity to share some details that may not be interesting from a historical point of view, but will likely brighten the story.
"When the pipe machine was inaugurated, a pompous blessing was organised. Only five guests were allowed to stay on the platform facing the control board, among them was the Archbishop of Caracas, the President of the company, the President of the group, who had come from France for the occasion, and the Director of the factory. That was without taking local customs into account. In the minutes immediately prior the ceremony, when the officials were in place, the invited crowd rushed over, climbed onto the platform, and packed as much as possible onto the small stage.
The start signal was sent, but nothing happened. The technicians had tested everything in the morning and everything was perfect. The people in charge started making a big hype about it, trying to figure out 'what the hell could be the source of the mess...' After a few minutes of panic, it appeared that the Archbishop, pushed by the invasive mob, pressed the emergency button...the solution was soon found".
Further details not closely related to the history of fibre-cement can also be reported:
-A little simple-minded employee gets his salary from the company, but nobody ever sees him except on paydays. His brother is a close co-worker of the minister who signs Canacit prefabricated house orders every year.
-Power outages are common, with the consequences one can imagine for machines and fresh fibre-cement paste. Explanation: Snakes, which are quite numerous in the area, climb posts like they used to do on trees. Once at the top, they twist around

high voltage cables, causing short circuits and falling to the ground, where they wiggle for a while. *(It is not a big story; This has been confirmed by several former Canacit employees).*
Sometimes the pumps stand still in the morning. When maintenance inquires about the problem, it seems that they were simply stolen...
Incidents did not prevent the company from supplying the pipes and sheets needed for the country's growth and even develop large self-supporting roofing sheets up to 9.20 meters in length (30.2ft), as seen at Caracas Airport. Picture from other times, far away from us.

08-059 Prefabricated country houses *08-060 Caracas Airport in the 1960s.*

In another field, it can be noted that relationships between workers and management have not always been at their best, as can be read in the following union leaflet from 1978. *(Original and translation next page).*

At the end of the 20th century I visited a factory belonging to Amanco Plycem group very close to Caracas. It seems to have disappeared, as has Tubacem-Canacit, which had been closely associated with Saint-Gobain. The present situation in Venezuela is not very favourable for an intense search.

SINDICATO DE TRABAJADORES DE ASBESTO, CEMENTO, SIMILARES Y CONEXOS DEL EDO. MIRAND/
S I T R A S B E S T O
AFILIADO A FETRACEMENTO
CASA SINDICAL: Calle Padre Arroyo No. 32 – El Rodeo – Ocumare del Tuy

ALERTA

COMUNICADO

El Sindicato hace del conocimiento a todos los Trabajadores de ASICA y TUBACEM que el día Martes 21 de Marzo de 1.978 la Empresa, en las personas de Michel Baur Jefe de Fabrica y Pinguet Director Tecnico, se dieron a la tarea de tratar de retirar al Sr. Vicente Palacios Secretario de Cultura y Propaganda de nuestro Sindicato, sabiendo que goza del Fuero Sindical que preves la Ley del Trabajo vigente en su Artículo N° 204. En ningún momento aceptamos las pretenciones y el abuso de estos Señores que piensan y se creen que estamos en los tiempos de la esclavitud.-

! NO A LA REPRESION PATRONAL¡¡¡

! NO A LOS QUE ABUSAN Y VIOLAN LA LEY DEL TRABAJO¡¡¡

! CONTRA LA ARREMETIDA PATRONAL UNIDAD DE LOS TRABAJADORES.

La Junta Directiva.

Communiqué

The union informs every worker of ASICA and TUBACEM that the Company, represented by Michel Bour, Plant Manager, and Bernard Pinguet, Technical Director, decided on Tuesday 21st of March 1978,to fire Mr. Vicente Palacio, Culture and Propaganda Secretary of our Union though they knew he benefited from the union Code of Laws, Article N°204 was included in the Labor Law. In any case, we can't accept that the above two people think and believe that we are still in the time of slavery.

NO TO MANAGERIAL CRACKDOWN

NO TO ABUSING AND VIOLATING LABOR LAW

AGAINST MANAGERIAL CRACKDOWN: UNIFIED WORKERS

Vietnam

Once again, one morning of 1995 or 1996, a phone call would take me to one of the countries that I had always dreamed of.
This time, the call did not come from a large international company, but from a young and small company in the nearby Deux-Sèvres district[183]. Although it did not specialise in fibre-cement it had sold, I wonder what a miracle, a flow-on machine brought to an equally young and small Vietnamese company. To make the machine run, they needed felts. With the help of the "Yellow pages" they found me and without much recognition we received an order for Vietnam, without any financial risk, as our French neighbour would pay for it. *(In the opinion of my boss, I rose again, without much merit.)*
Proud of this order and ready to enter the Vietnamese market, I would also like to see what the country looked like after all the years of war and used my contacts in Saint-Gobain to find a local agent in Hochiminh City, a town that I kept calling Saigon in my heart. Within a few months, we received fels orders from three other companies. Then I could make my dream a reality and on the occasion of one of my commercial visits to Thailand I extended my stay for the former French protectorate.
During my visit to the first customer, I was surprised to discover very small corrugated sheets. As far as I can remember, they were more or less 60 x 60 cm (23 x 23in). It became clear on my way back from the factory. The Vietnamese were not yet in the era "everyone-owns a car" . Bicycles were queens of the road along with mopeds, which were locally called motorcycles. Small corrugated sheets were perfectly adapted to racks. Cyclists rode with two or three of them on the front rack and two or three more on the rear one, when racks were clear of family members.
At the end of the 20th century, when I was selling felts in Vietnam, I had listed about ten production lines for asbestos-cement. In 2015, an NGO fighting against asbestos in the country claimed to have counted up to forty-one…"

"*In Vietnam, white asbestos is mostly used to produce asbestos-cement (AC) roofing sheets. Forty-one roofing sheet facilities nationwide can produce over 100 million*

183 Deux-Sèvres: rural French district with a "common border" with Charente, Ouest of France, that means one-hour drive from the company where I was then working.

sqm of AC roofing sheets annually, covering 60 percent of the demand, primarily in rural and mountainous areas due to their low prices and high level of durability".

A physician belonging to an organisation fighting for the asbestos ban, said: *"…a region where 85% of the houses are roofed with fibre-cement asbestos sheets and only 5% of the population have heard of the health hazard…"*

You simply understand the results of companies that operate the mines that are still active.

But not everything is so gloomy in Vietnam; VITD published on Marsh 3rd, 2013:

"We are the first company in Vietnam that succeeded to manufacture asbestos-less corrugated sheets with PVA, the best synthetic fibre to replace asbestos. Our products are made on modern automated lines, in conformity with the strictest norms, such as: Japan JIS A 5430:2001, Korea KS L5114/1998 and Vietnam TCVN:2000…Please let us know what kind of sheet is interesting you, we will send you an offer…"

The Website of Make-in-Vietnam is also very rich in technical information and may well deserve a visit.

And finally, the Vietnamese Government vows that asbestos will definitely be banned in 2023.

"Vietnam bans asbestos campaigns – The Vietnamese Prime Minister responds with an historic announcement of a deadline for banning asbestos in the construction sector…

"On January 16, 2018, the Vietnamese Prime Minister announced 2023 as the deadline for ending the use of asbestos in the construction sector. This was the first time the Prime Minister made such an announcement and it happened when he addressed the Ministry of Construction. The Prime Minister noted the previous 'obstruction' by the Ministry of Construction to push ahead with a ban. As a member of the Vietnam Ban Asbestos Network (VNBAN), the ban network is supported by the Union Aid Abroad APHEDA and directly quoted by the Prime Minister in his nationally televised speech…".

08-060...Corrugating device. Private archives.

Z for

Zimbabwe

Traditionally a large producer of asbestos, Zimbabwe closed its Mashaba and Shabanie mines in 2004. The decision created both relief and sorrow. Relief from the fact that the workers stopped going to the killing mine, sorrow because lack of work means poverty and misery.

In the second decade of our century, the country wonders and quarrels about the reopening of the mines. Investors from India seemed to be ready to offer money and to promise large purchase of the forthcoming asbestos. Many people refused and the government hesitated.

In November 2006, Dominique Baillard published a report on the situation based on RFI[184]. Here you can read:

"In order to defend its two asbestos mines, Zimbabwe implores South Africa to give up his project of embargo on imports of the disputed chrysotile fibre. A delegation of high-level representatives was sent to Pretoria to, once again, put asbestos health risk into perspective and mainly to insist on the economic consequences of such a decision. In 2005, the two mines of Shabanie and Mashaba had induced 40 million dollars income to the Harare government. That was a significant source of foreign currency in a beleaguered country. ZANU-PF, the ruling party, took control on mines two years ago by depriving of Zimbabwean nationality and expulsing the owner.

Since that time, he became a South-African citizen and he is claiming around that incomes resulting from the abhorred fibre are now used to refund an IMF credit...

...law men insisted most on their partners to consider the economic aspect of the embargo, saying that more or less 100,000 people are indirectly living from asbestos exports. The argumentation is worth any other. South Africa is ready to bring the ban into force from 2007, not only for sanitary reasons, but also under pressure of Everite's[185] lobby. The company which formerly produced the mineral is now the main support of its elimination".

---

[184] RFI: French International Broadcasting.

[185] Reminder: Everite South Africa is in no case associated with the French Everite which was quoted so many times in the book.

According to my recent search, the question of reopening the mines is still a topic, supporters and opponents arguing more than ever.
At the end of this "Countries Overview", let me quote a few lines published by Allied Market Research in February 2017:
"The global fibre-cement market is driven by factors such as rapid urbanization and industrialization in developing countries, booming construction industry, high efficiency of fibre-cement products, and ban on asbestos cement products. In addition, increased investment in infrastructure sector offers opportunities to the market players. However, lack of skilled labour in developing countries, hamper the growth of the global fibre-cement industry".
Finally, here is a non-exhaustive list of the most important holdings, groups and companies that are involved in the huge fibre-cement market according to Allied Market Research:
"Etex Group NV, James Hardie Industries PLC, Evonik Industries AG, Compagnie de Saint Gobain SA, Toray Industries Inc., CSR Limited, The Siam Cement Public Company Ltd, Nichiha Corporation, Plycem Corporation, Cembrit Holding A/S.
But also: Elementia, Marley Eternit Ltd, Thai Olympic Fibre-Cement Co. Ltd, Mahaphant Fibre-Cement Co. Ltd., Everest Industries Ltd., Swisspearl, Equitone, Allura, Beijing Hocreboard Building Materials Co. Ltd., Fry Reglet."
Indeed, I am sorry for the companies that I may have forgotten or offended by inadequate words during the so difficult task of telling the worldwide history of the fibre-cement industry.

Epilogue

Everything you have just read relates to facts that may seem new to some of us and distant to others.

All over the world, from product to product, from machine to machine, from country to country, with the help of states to form strategic alliances, the industry has built factories ...

Born in Austria almost 120 years ago, developed, globalised, often depressed, ill, back on its feet, the "Fibre(s)cement" industry will certainly surprise us again.

Generations have strived, suffered, created for their own lives or improved that of their customers, who brought them to life, to roof them, to decorate them, and to see that their water got to the family tap ...

Investors have increased their wealth, engineers have done research, workers have worked hard, they have even lost their lives at times. You will ask what for.

I think a lot! Isn't it difficult to live without a roof? Without running water from the tap? Without sewage drainage? Do not pretty colours help our quality of life?

In the past twenty or thirty years a new page has been written on the fibre-cement industry. New fulfilments, perhaps still incomplete or uncertain, seemed unthinkable at the 1991 conference in Kuala Lumpur.

Let us admit that the various studies of asbestos-free fibre-cement have not yet fully proven their long-term survival and that improvement is still foreseeable. In this industry, however, as can be seen on the first page of the book, it often takes decades to achieve a perfect degree of maturity of the manufacturing process and the resulting product.

Of course, fibre-cement is not the only material that offers solutions, but it is endlessly replicable and the most reliable prospective studies promise flourishing days for years to come.

A simple look at the new factories in countries with the greatest need for housing and infrastructure will convince everyone, and if not, a visit to Allied Market Research, where you can read:

*"**Global Fibre-Cement Market** generated revenue of $12,336 million in 2014, and is expected to garner $18,888 million by 2022, registering a CAGR of 5.8% from 2016 to 2022".*

In June 2019, another prospective study published a report announcing a very similar future.

Global report on fibre-cement sheets in main countries 2019-2023.

Press Release

Fibre-Cement Board Market Size 2019-2023 Capacity, Production (K Units), Revenue (Million USD)

Published: June 3, 2019 3:15 a.m. ET

"June 03, 2019 (The Express wire via COMTEX) – Fibre-Cement Board Market Report focuses on the key global manufacturers, to define, describe and analyse the sales volume, value, market share, market competition landscape, SWOT analysis and development plans.

"A complete analysis report created through intensive primary research (inputs from business specialists, companies, stakeholders) and secondary research, the report aims to present the analysis of Global Fibre-Cement Board Market 2018 to 2023. The global fibre-cement board market has been analysed by product type and application as well as by region (North America, Europe, Asia-Pacific, rest of the World) and by country (U.S, Canada, U.K, Germany, China, India, Japan and Brazil) for the historic time period and also the forecast period of 2018-2023.

"Over the past few years, the global fibre-cement board market has been witnessing an enormous growth and is expected to take its dominant position in the fibre-cement board industry. Moreover, the report conjointly classifies and analyses the evolving trends as well as the major drivers, challenges and opportunities in the global fibre-cement board market..."

Thanks

A special warm thanks to:

Alain Sabouraud whose knowledge, memory and patience were more than bearable requested,

Fernando de Aragón Ó, for the same reason

Roberto	Abujder	Bolivia	
Samuel	Bouré	France	
Orlando	Barrial i Jové	Spain	
Cécile	Belot	France	
Robert	Bessiron	France	
Charles	Biaggi	France	
Michel	Bour	Colombia	
Raymond	Campestrini	France	
Sophie	Chavignon	France	
John	Cottier	South Africa	
Fernando	De Aragón Ó	Mexico	
Olivier	Delas	France	
Yann	Deret	France	
Marcel	Descombe	France	deceased

Pierre	Dieudonné	France	deceased
Vincent	Favre	France	
Carlos	Freitas	Portugal	
Daniel	Friedman	USA	
Oliver	Glendenning	UK	
Michel	Gonzalez	France	
Alexandra	Kern	Switzerland	
Marc	Lamour	France	
Annie	Lamontagne	Canada	
Laure	Lanterie	France	
Bernard	Mallet	France	
Christian	Manant	France	
Joseph	Milewski	France	deceased
Ginette	Milewski	France	
Bernard	Montagut	?	
Denis	Peaucelle	France	
Pierre	Picavet	France	
Bernard	Pinguet	France	
Fabienne	Planchenot	France	
Francis	Queva	France	
Elisabeth	Rauchenzauner	Austria	
Emilie	Roulland	France	

Thanks

Manuel	Rubio Morano	Spain
Robert	Ruers	Netherlands
Alain	Sabouraud	France
Gérard	Sandret	France
Anonymous person	Sinoma Industry	China
Pierrette	Theroux	Canada
José María	Txecma Romero	Spain
Francisco	Vicke	México
Dietz	Torsten	Germany